Lecture Notes in Computer Science　　11791

More information about this series at http://www.springer.com/series/7412

Kilian M. Pohl · Wesley K. Thompson ·
Ehsan Adeli · Marius George Linguraru (Eds.)

Adolescent Brain Cognitive Development Neurocognitive Prediction

First Challenge, ABCD-NP 2019
Held in Conjunction with MICCAI 2019
Shenzhen, China, October 13, 2019
Proceedings

 Springer

Editors
Kilian M. Pohl ⓘ
Stanford University
Stanford, CA, USA

Ehsan Adeli ⓘ
Stanford University
Stanford, CA, USA

Wesley K. Thompson
University of California, San Diego
La Jolla, CA, USA

Marius George Linguraru ⓘ
Children's National Health System
Washington, DC, USA

ISSN 0302-9743   ISSN 1611-3349 (electronic)
Lecture Notes in Computer Science
ISBN 978-3-030-31900-7   ISBN 978-3-030-31901-4 (eBook)
https://doi.org/10.1007/978-3-030-31901-4

LNCS Sublibrary: SL6 – Image Processing, Computer Vision, Pattern Recognition, and Graphics

This Springer imprint is published by the registered company Springer Nature Switzerland AG
The registered company address is: Gewerbestrasse 11, 6330 Cham, Switzerland

Preface

The ABCD Neurocognitive Prediction Challenge (2019 ABCD-NP-Challenge; https://sibis.sri.com/abcd-np-challenge) invited researchers to submit methods for predicting fluid intelligence from T1-weighted MRI of 8669 children (age 9–10 years) recruited by the Adolescent Brain Cognitive Development Study (ABCD) study—the largest long-term study of brain development and child health in the United States to date. The first ABCD-NP-Challenge was organized in conjunction with the 22nd International Conference on Medical Image Computing and Computer Assisted Intervention (MICCAI) held in October 2019 in Shenzhen, China. In total, 124 teams registered for the challenge. There were no limits or restrictions on team members as long as the team complied with the National Institute of Mental Health (NIMH) Data Archive Data Use Certification of the ABCD project and members of the team were not from labs associated with the ABCD study (https://abcdstudy.org/). Each team was instructed to use the NIMH Data Archive (NDA) portal to download the T1-weighted MRI, which the organizers had skull-stripped and affinely aligned to the SRI 24 atlas. In addition, the challenge organizers provided brain parcellations defined by the atlas and a CSV file with the corresponding volume measurements of each region of interest (ROI). Information about the ABCD Data Repository can be found at https://nda.nih.gov/abcd/about.

The NDA portal also provided the residual fluid intelligence scores of 4154 subjects for training (3739 samples) and validation (415 samples). Fluid intelligence is a major component in determining general intelligence. Determining the neural mechanisms underlying general intelligence is fundamental to understanding cognitive development, how this relates to real-world health outcomes, and how interventions (education, environment) might improve outcomes through adolescence and into adulthood. The fluid intelligence scores recorded by the ABCD study were measured via the NIH Toolbox Neurocognition battery. The scores provided by the challenge organizers were pre-residualized on data collection site, sociodemographic variables, and brain volume. The R code for computing the residual scores was accessible through the challenge website. The residual fluid intelligence scores of the 4515 subjects used to test each method were not released but had to be predicted based on the provided T1-weighted MRI. The corresponding raw fluid intelligence scores and demographic factors were first made accessible to the public via the ABCD Data Release 2.0, which was released after the submission deadline (March 24, 2019) of the challenge.

There were 29 submissions, of which 24 were accepted to the challenge after passing a single-blinded review. An eligible submission consisted of a CSV file containing the predictions of the fluid intelligence based only on the provided T1-weighted MRIs of at least 99% of the 4402 test subjects, the source code generating those predictions, and a manuscript describing the method and findings. The document needed to clearly describe the data used for prediction, the method, and findings including the prediction error during training and validation. Authors submitting multiple manuscripts needed to describe methods and results that were different from each other and from previously published material. With these criteria, manuscripts of some submissions were merged resulting in a total of 21 papers, which are included in this book regardless of the ranking in the challenge. Each paper is described in one chapter and includes detailed implementations steps, analysis of the results, and comparison with baseline methods.

Contestants were ranked separately on the validation data set and on the test data sets. For each data set, the organizers computed the mean squared error (MSE) between their predicted scores and the pre-residual fluid intelligence scores according to publicly available R code. The error of missing predictions was the largest MSE from among the set of values produced by the same algorithm on the subjects in the dataset. Overall, 19 submissions (out of 24) were better than a naïve predictor, i.e., the mean intelligence score based on the training data. These results revealed that the 2019 ABCD-NP-Challenge was grand. Structural T1-weighted MRI should contain more information about fluid intelligence as at that age intelligence is not yet considered a result of education and thus mostly associated with family history, including genetics. Of the MRI modalities acquired by ABCD, T1-weighted MRI modalities most closely linked to genetics.

August 2019

Kilian M. Pohl
Wesley K. Thompson
Ehsan Adeli
Marius George Linguraru

Organization

Chairs

Kilian M. Pohl   Stanford University and SRI International, USA
Wesley K. Thompson   University of California San Diego, USA

Co-chairs

Ehsan Adeli   Stanford University, USA
Bennett K. Landman   Vanderbilt University, USA
Marius George Linguraru Children's National Health System and George
                            Washington University, USA
Susan Tapert   University of California San Diego, USA

Contents

A Combined Deep Learning-Gradient Boosting Machine Framework for Fluid Intelligence Prediction

Yeeleng S. Vang$^{(\boxtimes)}$, Yingxin Cao, and Xiaohui Xie

University of California, Irvine, CA 92697, USA
{ysvang,yingxic4,xhx}@uci.edu

Abstract. The ABCD Neurocognitive Prediction Challenge is a community driven competition asking competitors to develop algorithms to predict fluid intelligence score from T1-w MRIs. In this work, we propose a deep learning combined with gradient boosting machine framework to solve this task. We train a convolutional neural network to compress the high dimensional MRI data and learn meaningful image features by predicting the 123 continuous-valued derived data provided with each MRI. These extracted features are then used to train a gradient boosting machine that predicts the residualized fluid intelligence score. Our approach achieved mean square error (MSE) scores of 18.4374, 68.7868, and 96.1806 for the training, validation, and test set respectively.

Keywords: Fluid intelligence · Deep neural network · Machine learning

1 Introduction

The Adolescent Brain Cognitive Development (ABCD) study [2] is the largest long-term study of brain development and child health in the United States. Its stated goal is to determine how childhood experiences such as videogames, social media, sports, etc. along with the child's changing biology affects brain development. Understanding of brain development during the adolescent period is "necessary to permit the distinction between premorbid vulnerabilities and consequences of behaviors such as substance use" [14]. In this endeavor, leaders of the study organized the ABCD Neurocognitive Prediction Challenge (ABCD-NP-Challenge 2019) [1] and invited teams to make predictions about fluid intelligence from T1-w magnetic resonance images (MRI). In psychology and neuroscience, the study of general intelligence often revolves around the concepts of fluid intelligence and crystallized intelligence. Fluid intelligence is defined as the ability to reason and to solve new problems independent of previously acquired knowledge [11] whereas crystallized intelligence is the ability to use skill and

---

Y. S. Vang and Y. Cao—Equal contribution.

© Springer Nature Switzerland AG 2019
K. M. Pohl et al. (Eds.): ABCD-NP 2019, LNCS 11791, pp. 1–8, 2019.
https://doi.org/10.1007/978-3-030-31901-4_1

previous knowledge. General intelligence is usually quantified by tests such as the Cattell Culture Fair Intelligence Test or Raven's Progressive Matrices.

This paper represents the authors' entry to the ABCD-NP-Challenge 2019 competition. We propose a Convolutional Neural Networks (CNN) combined with gradient boosting machine (GBM) framework for the task of fluid intelligence prediction from 3D T1-w MRIs. Our method combines the state-of-the-art approach of deep learning to find a good, non-linear compression of the high dimensional 3D MRI data and uses the superior performance of GBM to learn an ensemble regression model.

2   Related Work

The study of fluid intelligence has traditionally been more concern with trying to identify the underlying mechanism responsible for cognitive ability. In [5,9], experiments were conducted to investigate factors and mechanisms that influence fluid intelligence with both suggesting that the lateral prefrontal cortex may play a critical role in controlling processes central to human intelligence. More recently, MRIs have been shown to contain useful structural information with strong correlation to fluid intelligence [6]. In the most related work to the subject of this paper, fluid intelligence was predicted directly from MRI data using support vector regressor obtaining an average correlation coefficient of 0.684 and average root mean square error of 9.166 [19], however the dataset consisted of only 164 subjects.

In recent years, deep learning [13] method has emerged as state-of-the-art solutions to many problems spanning various domain such as natural language processing, bioinformatics, and especially computer vision. Since winning the ImageNet competition in 2012 [12], CNN, a type of deep learning model, have been the defacto tool for analyzing image data. Its impact in the biomedical image domain has been nothing short of extraordinary [17]. Many disease, such as cancer, that are detectable and segmentable by radiologists studying brain MRI can now be automatically performed by deep learning algorithms [3,10] with performances now comparable to many experts [18,20].

3   Model

Our model pipeline consists of two parts. As these 3D MRIs are of very high dimensions, we first train a CNN model to learn a data compression scheme that best preserves meaningful features of the original image data. Using this CNN, we feature extract a compressed version of the original input image to better utilize limited computer memory. These compressed images are used as surrogates for the original images and is used to train the GBM to learn a prediction model. These two steps are explain in further details in the following sections.

3.1   Feature Extractor

The original T1-w MRIs are much too high dimensional to be used with many modern machine learning algorithms with more than a few samples at a time due to computing memory limitations. This necessitates first compressing these original MRIs to a more compact representation while preserving meaningful features before learning a regression model. Another motivation for using a compact representation is to reduce potential overfitting. In our proposed framework, we use a fully convolutional network (FCN) to encode the raw MRIs to a compressed form. It is fully convolutional because we only use the convolution and pooling operations. The structure of the CNN is shown in Fig. 1.

The dual-channel input to our FCN consists of the original T1-w MRI and segmented brain MRI. These two images are cropped into 192^3-voxel cubes and passed through our FCN resulting in a real-valued vector output of length 123. The length of this one-dimensional vector is design to match the number of derived, continuous covariates for each MRI [15]. These derived data are discussed in more detail in the data section. We formulate the training of the feature extractor as a regression problem calculating the mean square error (MSE) between predicted output and the continuous covariates as the loss function. The motivation to use this approach is due to the fact that deep learning models have trouble predicting a single continuous value when the input is of such high dimensions. As the 123 continuous covariates are unique for each subjects, the thought is that our model will learn better features for downstream task. We also design the FCN in an encoder fashion to permit easy extracting of features at different image scales.

The model is trained for 15 epochs with a learning rate of 0.01, momentum of 0.9, and batch size of 4. The stochastic gradient descent with momentum optimizer is used. The model is trained on both the training and test datasets and validated on the validation set. The epoch with the lowest validation set MSE score was used to extract features. We extracted features at both the $6 \times 6 \times 6$ and $3 \times 3 \times 3$ scales. We found empirically that the $6 \times 6 \times 6$ scale produced better results in the latter fluid intelligence prediction part.

3.2   Gradient Boosting Machine

Following feature extraction, the extracted images are regressed upon using gradient boosting machine (GBM) [7]. GBM is a boosting method that obtains a strong predictor by ensembling many weak predictors. It iteratively improves its current model by adding a new estimator fitted to the residual of the current model and true labels. Specifically, GBM learns a functional mapping $\hat{y} = F(x; \beta)$ from data $\{x_i, y_i\}_{i=1}^N$ that minimizes some loss function $L(y, F(x; \beta))$. More concretely, $F(x)$ takes the form of $F(x) = \sum_{m=0}^M \rho_m f(x; \tau_m)$ where ρ is the weight, f is the weak learner, and τ is the parameter set. Then β consists of the parameter sets $\{\rho_m, \tau_m\}$. These parameter sets are learned in the following stage-wise greedy process:

{1} set an initial estimator $f_0(x)$.

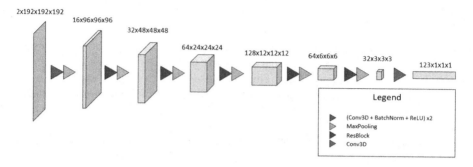

Fig. 1. Illustration of feature extractor model. This is a fully convolutional network designed in an encoder architecture for easy access to data compression at different image resolutions.

{2} for $m \in 1, 2, \ldots, M$

$$(\rho_m, \tau_m) = \arg \min_{\rho, \tau} \sum_{i=1}^{n} L(y_i, F_{m-1}(x_i) + \rho f(x_i; \tau)) \tag{1}$$

Step {2} is approximated by GBM in the following two steps:
First, learn τ by:

$$\tau_m = \arg \min \sum_{i=1}^{n} (g_{im} - f(x_i; \tau))^2 \tag{2}$$

where $g_{im} = -\left[\dfrac{\partial L(y_i, F(x_i))}{\partial F(x_i)} \right]_{F(x) = F_{m-1}(x)}$
Second, learn ρ by:

$$\rho_m = \arg \min_{\rho} \sum_{i=1}^{n} L(y_i, (F_{m-1}(x_i) + \rho f(x_i; \tau_m))$$

Finally it updates $F_m(x) = F_{m-1}(x) + \rho f(x; \tau)$. To control overfitting, shrinking is introduced to give the following form of the update equation: $F_m(x) = F_{m-1}(x) + \gamma \rho f(x; \tau)$, where $0 \leq \gamma \leq 1$. The performance of GBM is improved by random subsampling the training data to fit each new estimator [8].

In this project, we use the XGBoost [4] implementation of GBM for fast GPU support using trees. The GBM model requires a number of hyperparameters to be set. We perform a two stage grid search to find the best combination of hyperparameters. The first stage involves searching over a coarse grid to obtain the general vicinity of the optimal hyperparameters. The second stage involves searching over a fine grid around the best hyperparameter from the first stage. The following hyperparameters are found to give the best validation set mean square error score: Learning rate = 0.006, number of trees = 1000, depth of tree = 7.

To minimize overfitting, we use elastic net regularization [21] which adds the LASSO and RIDGE regularization to Eqs. (1) and (2). Specifically, $\frac{1}{2}\lambda\sum_{m=1}^{M}\rho^2 + \alpha\sum_{m=1}^{M}|\rho|$ is added to Eq. (1) and $\frac{1}{2}\lambda\sum_{m=1}^{M}\tau^2 + \alpha\sum_{m=1}^{M}|\tau|$ is added to Eq. (2). The elastic net regularization parameters are tuned against the validation dataset and set as $\lambda = 1.05$ and $\alpha = 0.1$ to achieve the results of Table 1.

4 Experiment

4.1 Data

3D T1-w MRIs were pre-processed and provided by the challenge organizer. The pre-processing steps involved first creating brain masks from a series of standard brain extraction softwares including FSL BET, AFNI 3dSkullStrip, FreeSurer mri_gcut, and Robust Brain Extraction (ROBEX). The final brain mask was obtained by taking majority voting across these resulting masks. This final brain mask was used to perform bias correction and the extracted brain was segmented into brain tissue (gray matter, white matter, and cerebrospional fluid (CSF)) using Atropos. Finally the skull-stripped brain and segmented brain images were registered affinely to the SRI24 atlas [15]. Figure 2 shows typical example of the pre-processed, skullstripped T1-w MRI and segmented brain MRI provided by the organizers.

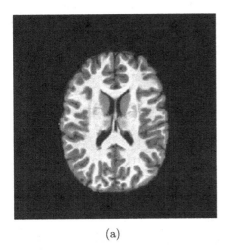
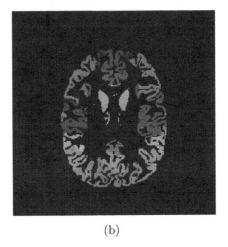

(a)   (b)

Fig. 2. (a) Example of pre-processed T1-w brain MRI. (b) Example of the white matter, gray matter, and CSF segmented brain MRI of the same T1-w image.

The fluid intelligence scores were pre-residualized by the challenge organizer. It involved fitting a linear regression model to fluid intelligence as a function of the following covariates: brain volume, data collection site, age at baseline, sex at birth, race/ethnicity, highest parental education, parental income, and parental marital status. Subjects in the ABCD NDA Release 1.1 dataset missing any value was excluded from the training and validation set. Once the linear regression model was fitted, fluid intelligence residual for each patients were calculated and constitutes the value to predict for this competition.

In addition to the provided 3D T1-w MRI, segmented brain MRI, and the fluid intelligence residual scores, derived data [15] was provided for each patient. These derived data consisted of 123 continuous-valued, volumetric data for different gray and white matter tissues such as "right precentral gyrus gray matter volume" and "corpus callosum white matter volume". In all, the training set consists of 3739 patients, validation set consists of 415 patient, and test set consists of 4402 patients.

4.2   Evaluation Criteria

The competition evaluation criteria is based on the mean square error (MSE) score between predicted and true residualized fluid intelligence score. This is calculated as the following equation:

$$MSE = \frac{1}{N} \sum_{i=1}^{N} (y - \hat{y})^2$$

where N is the total number of subjects in each data fold, y is the true residualized fluid intelligence score, $\hat{y}$ is the predicted score from our model. The organizers provided an evaluation script written in the R language to perform this evaluation for the training and validation set for consistency between competitors. Ranking is provided separately for both validation set performance and test set performance.

Table 1. Proposed network MSE performance

	Train set	Validation set	Test set
BrainHackWAW	—	**67.3891**	92.9277
MLPsych	—	68.6100	95.6304
CNN+GBM (our method)	18.4374	68.7868	96.1806
BIGS2	—	69.3861	93.1559
UCL CMIC	—	69.7204	**92.1298**

4.3   Results

The result of our algorithm is reported in Table 1 along with the other top four performing teams based on MSE ranking of the validation set performance at the end of the competition. Our approach achieved a third place finish on the validation set. The best performing team based on the test set was 4.3% better than our results.

Table 2 shows the ablation study proving the value of incorporating MRI date for fluid intelligence prediction. The baseline "derived data+GBM" learns a GBM regression model on the derived data directly. Using our proposed method incorporating T1-w MRI data reduces training set MSE by 25.5% and validation set MSE by 5.4%.

Table 2. Ablation study

	Training set MSE	Validation set MSE
Derived data+GBM	24.7683	72.7308
CNN+GBM	18.4374	68.7868

5   Conclusion

In this paper we propose a combination deep learning plus gradient boosting machine framework for the task of fluid intelligence prediction based on T1-w MRI data. Specifically we design a fully convolutional network to perform data compression/feature extraction. Using these extracted features, we employ gradient boosting machine to learn an ensemble model for regression. Our model achieved a third place finish based on the validation set ranking.

For future directions, we would like to investigate different encoding scheme such as autoencoders to perform the feature extraction step. In theory, autoencoders are able to find the optimal compression scheme which may improve the downstream regression task. About the regression model, instead of using elastic net as the regularization technique, we can consider replacing them with the more recently developed dropout [16] to minimize the gap in training set and validation set MSE performance for better generalization.

References

1. ABCD neurocognitive prediction challenge. https://sibis.sri.com/abcd-np-challenge/. Accessed 13 Mar 2019
2. Adolescent brain cognitive development study. https://abcdstudy.org/. Accessed 13 Mar 2019
3. Akkus, Z., Galimzianova, A., Hoogi, A., Rubin, D.L., Erickson, B.J.: Deep learning for brain MRI segmentation: state of the art and future directions. J. Digit. Imaging **30**(4), 449–459 (2017)
4. Chen, T., Guestrin, C.: XGBoost: a scalable tree boosting system. In: Proceedings of the 22nd ACM SIGKDD International Conference on Knowledge Discovery and Data Mining, pp. 785–794. ACM (2016)

5. Cole, M.W., Yarkoni, T., Repovš, G., Anticevic, A., Braver, T.S.: Global connectivity of prefrontal cortex predicts cognitive control and intelligence. J. Neurosci. **32**(26), 8988–8999 (2012)
6. Colom, R., et al.: Gray matter correlates of fluid, crystallized, and spatial intelligence: testing the P-FIT model. Intelligence **37**(2), 124–135 (2009). https://doi.org/10.1016/J.INTELL.2008.07.007. https://www.sciencedirect.com/science/article/pii/S0160289608000925
7. Friedman, J.H.: Greedy function approximation: a gradient boosting machine. Ann. Stat. **29**, 1189–1232 (2001)
8. Friedman, J.H.: Stochastic gradient boosting. Comput. Stat. Data Anal. **38**(4), 367–378 (2002)
9. Gray, J.R., Chabris, C.F., Braver, T.S.: Neural mechanisms of general fluid intelligence. Nature Neurosci. (2003). https://doi.org/10.1038/nn1014
10. Havaei, M., et al.: Brain tumor segmentation with deep neural networks. Med. Image Anal. **35**, 18–31 (2017)
11. Jaeggi, S.M., Buschkuehl, M., Jonides, J., Perrig, W.J.: Improving fluid intelligence with training on working memory. Proc. Natl. Acad. Sci. **105**(19), 6829–6833 (2008)
12. Krizhevsky, A., Sutskever, I., Hinton, G.E.: ImageNet classification with deep convolutional neural networks. In: Advances in Neural Information Processing Systems, pp. 1097–1105 (2012)
13. LeCun, Y., Bengio, Y., Hinton, G.: Deep learning. Nature **521**(7553), 436 (2015)
14. Luciana, M., et al.: Adolescent neurocognitive development and impacts of substance use: overview of the adolescent brain cognitive development (ABCD) baseline neurocognition battery. Dev. Cogn. Neurosci. **32**, 67–79 (2018)
15. Pfefferbaum, A., et al.: Altered brain developmental trajectories in adolescents after initiating drinking. Am. J. Psychiatry **175**(4), 370–380 (2017)
16. Rashmi, K.V., Gilad-Bachrach, R.: DART: dropouts meet multiple additive regression trees. In: AISTATS, pp. 489–497 (2015)
17. Shen, D., Wu, G., Suk, H.I.: Deep learning in medical image analysis. Ann. Rev. Biomed. Eng. **19**, 221–248 (2017)
18. Tang, H., Kim, D.R., Xie, X.: Automated pulmonary nodule detection using 3D deep convolutional neural networks. In: 2018 IEEE 15th International Symposium on Biomedical Imaging, ISBI 2018, pp. 523–526. IEEE (2018)
19. Wang, L., Wee, C.Y., Suk, H.I., Tang, X., Shen, D.: MRI-based intelligence quotient (IQ) estimation with sparse learning. PloS one **10**(3), e0117295 (2015)
20. Zhu, W., et al.: AnatomyNet: deep learning for fast and fully automated whole-volume segmentation of head and neck anatomy. Med. Phys. **46**, 579–589 (2018)
21. Zou, H., Hastie, T.: Regularization and variable selection via the elastic net. J. Roy. Stat. Soc.: Ser. B (Stat. Methodol.) **67**(2), 301–320 (2005)

Predicting Fluid Intelligence of Children Using T1-Weighted MR Images and a StackNet

Po-Yu Kao[1]([✉]) [iD], Angela Zhang[1], Michael Goebel[1], Jefferson W. Chen[2], and B. S. Manjunath[1]([✉])

[1] University of California, Santa Barbara, CA, USA
{poyu_kao,manj}@ucsb.edu
[2] University of California, Irvine, CA, USA

Abstract. In this work, we utilize T1-weighted MR images and Stack-Net to predict fluid intelligence in adolescents. Our framework includes feature extraction, feature normalization, feature denoising, feature selection, training a StackNet, and predicting fluid intelligence. The extracted feature is the distribution of different brain tissues in different brain parcellation regions. The proposed StackNet consists of three layers and 11 models. Each layer uses the predictions from all previous layers including the input layer. The proposed StackNet is tested on a public benchmark Adolescent Brain Cognitive Development Neurocognitive Prediction Challenge 2019 and achieves a mean squared error of 82.42 on the combined training and validation set with 10-fold cross-validation. The proposed StackNet achieves a mean squared error of 94.25 on the testing data. The source code is available on GitHub (https://github.com/UCSB-VRL/ABCD-MICCAI2019).

Keywords: T1-weighted MRI · Fluid intelligence (Gf) · Machine learning · StackNet

1 Introduction

Fluid intelligence (Gf) refers to the ability to reason and to solve new problems independently of previously acquired knowledge. Gf is critical for a wide variety of cognitive tasks, and it is considered one of the most important factors in learning. Moreover, Gf is closely related to professional and educational success, especially in complex and demanding environments [7]. The ABCD Neurocognitive Prediction Challenge (ABCD-NP-Challenge 2019) provides 8556 subjects, age 9–10 years, with T1-weighted MR images and fluid intelligence which is withheld for testing subjects. The motivation of the ABCD-NP-Challenge 2019 is to discover the relationship between the brain and behavioral measures by leveraging the modern machine learning methods.

A few recent studies use structural MR images to predict fluid intelligence. Paul et al. [13] demonstrated that brain volume is correlated with quantitative

© Springer Nature Switzerland AG 2019
K. M. Pohl et al. (Eds.): ABCD-NP 2019, LNCS 11791, pp. 9–16, 2019.
https://doi.org/10.1007/978-3-030-31901-4_2

reasoning and working memory. Wang et al. [19] proposed a novel framework for the estimation of a subject's intelligence quotient score with sparse learning based on the neuroimaging features. In this work, we utilize the T1-weighted MR images of adolescents to predict their fluid intelligence with a StackNet. While whole brain volumes have been examined in relation to aspects of intelligence, to our knowledge there has been no previous work which examines the predictive ability of whole brain parcellation distributions for fluid intelligence. The main contributions of our work are two-fold: (1) to predict pre-residualized fluid intelligence based on parcellation volume distributions, and (2) to show the significance of the volume of each region on the overall prediction.

2   Materials and Methods

2.1   Dataset

The Adolescent Brain Cognitive Development Neurocognitive Prediction Challenge (ABCD-NP-Challenge 2019) [4,6,8,15,18] provides data for 3739 training subjects, 415 validation subjects and 4402 testing subjects (age 9–10 years). MR-T1 image is given for each subject, but the fluid intelligence scores are only provided for the training and validation subjects. MR-T1 images are distributed after skull-stripped and registered to the SRI 24 atlas [16] of voxel dimension $240 \times 240 \times 240$. In addition to the MR-T1 images, the distributions of gray matter, white matter, and cerebrospinal fluid in different regions of interest according to the SRI 24 atlas are also provided for all subjects. The fluid intelligence scores are pre-residualized on data collection site, sociodemographic variables and brain volume. The provided scores should, therefore, represent differences in Gf not due to these known factors.

2.2   StackNet Design

StackNet [10] is a computational, scalable and analytical framework that resembles a feed-forward neural network. It uses Wolpert's stacked generalization [20] in multiple levels to improve the accuracy of classifier or reduce the error of regressor. In contrast to the backward propagation used by feed-forward neural networks during the training phase, StackNet is built iteratively one layer at a time (using stacked generalization), with each layer using the final target as its target.

There are two different modes of StackNet: (i) each layer directly uses the predictions from only one previous layer, and (ii) each layer uses the predictions from all previous layers including the input layer that is called restacking mode. StackNet is usually better than the best single model contained in each first layer. However, its ability to perform well still relies on a mix of strong and diverse single models in order to get the best out of this meta-modeling methodology.

We adapt the StackNet architecture for our problem based on the following ideas: (i) including more models which have similar prediction performance, (ii) having a linear model in each layer (iii) placing models with better performance on a higher layer, and (iv) increasing the diversity in each layer. The resulting StackNet, shown in Fig. 1, consists of three layers and 11 models. These models include one Bayesian ridge regressor [9], four random forest regressors [1], three extra-trees regressors [5], one gradient boosting regressor [3], one kernel ridge regressor [12], and one ridge regressor. The first layer has one linear regressor and five ensemble-based regressors, the second layer contains one linear regressor and two ensemble-based regressors, and the third layer only has one linear regressor. Each layer uses the predictions from all previous layers including the input layer.

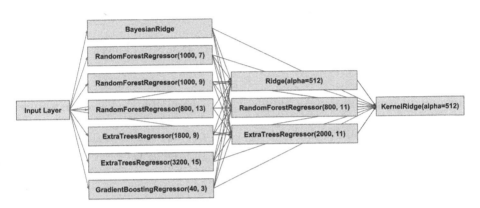

Fig. 1. The architecture of proposed StackNet. For the ensemble-based regressor, the number of trees and the maximum depth of each tree are indicated in the first and second number, respectively.

2.3   Predicting Gf Using Structural MR Images and StackNet

Figure 2 shows the framework of predicting the fluid intelligence scores using MR-T1 images and a StackNet. The framework is implemented with the scikit-learn [2,14] Python library. In the training phase, features are extracted from the MR-T1 images of training and validation subjects. We then apply normalization and feature selection on the extracted features. In the end, these pre-processed features are used to train the StackNet in Fig. 1. In the testing phase, features are extracted from the MR-T1 images of testing subjects, and the same feature pre-processing factors are applied to these extracted features. Thereafter, the pre-processed features are used with the trained StackNet to predict the fluid intelligence of the testing subjects. Details of each step are described below.

The ABCD-NP-Challenge 2019 data includes pre-computed 122-dimension feature that characterizes the volume of brain tissues, i.e., gray matter, white matter, and cerebrospinal fluid, parcellated into SRI-24 [16] regions. The feature extracted for each subject is defined as $f_i(j)$ where i is the index of subject and $j \in \{1, 2, \dots, 122\}$ is the index of feature dimension.

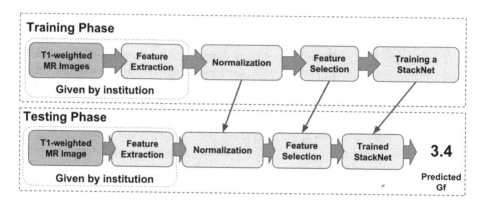

Fig. 2. The framework of predicting fluid intelligence using MR-T1 images and a Stack-Net.

Normalization: We apply a standard score normalization on each feature dimension, $\overline{f}_i(j) = (f_i(j) - \mu(j))/\sigma(j)$ where i is the index of subject, j is the index of feature dimension, and $\overline{f}_i(j)$ and $f_i(j)$ are the normalized and raw feature dimension j of subject i, respectively. $\mu(j)$ and $\sigma(j)$ are the mean and standard deviation of feature dimension j, respectively.

Feature Selection: Feature selection consists of three steps: (i) reducing the noise of data and generating an accurate representation of data through principal component analysis (PCA) [17] with maximum-likelihood estimator [11] (ii) removing the feature dimensions with the low variance between subjects, and (iii) selecting 24 feature dimensions with the highest correlations to the ground-truth Gf scores through univariate linear regression tests. Thereafter, the feature dimensions shrink from 120 to 24.

Training a StackNet: Because the mean of the pre-residualized fluid intelligence for the training dataset ($\mu = 0.05, \sigma = 9.26$) and validation dataset ($\mu = -0.5, \sigma = 8.46$) are quite different, we combine these two datasets ($\mu = 0, \sigma = 9.19$) for hyperparameter optimization and training a StackNet.

Predicting Fluid Intelligence: In the testing phase, we first apply the same pre-processing factors used in the training phase to the extracted features of testing subjects. We then use the trained StackNet with these pre-processed features to predict the fluid intelligence scores of testing subjects.

Evaluation Metric: The mean squared error (MSE) is used to calculate the error between the predicted Gf scores and the corresponding ground-truth Gf scores.

2.4 Computing Feature Importance

We would like to discover the correlation between the Gf score and the brain tissue volume in a region. Thus, we compute the importance of each feature dimension, and higher importance represents a higher correlation. However, after feature selection, the original data space of dimension 122 is projected and reduced to a new space of dimension 24. In this new space, we first compute the importance of each feature dimension and then backward propagate it to the original data space of dimension 122. The details are explained as follows.

After dimensionality reduction, we obtain the individual correlations between the remaining 24 feature dimensions and the ground-truth Gf scores. These correlations are first converted to F values and then normalized w.r.t. the total F values of feature dimensions, i.e., $\overline{F}_1 + \overline{F}_2 + ... + \overline{F}_{24} = 1$, where $\overline{F}_k$ is the normalized F value of feature dimension $k \in \{1, 2, ..., 24\}$. These normalized F values are used to build a normalized F vector as $\boldsymbol{F}_{1\times24} = [\overline{F}_1, \overline{F}_2, ..., \overline{F}_{24}]$. We then use the corresponding eigenvectors and eigenvalues from the PCA transformation to build two matrices, $\boldsymbol{U}_{122\times24} = [\vec{u}_1, \vec{u}_2, ..., \vec{u}_{24}]$ and $\boldsymbol{\Lambda}_{1\times24} = [\lambda_1, \lambda_2, ..., \lambda_{24}]$, where $\vec{u}_k$ and λ_k are the corresponding eigenvector and eigenvalue for $\overline{F}_k$, respectively. The dimension of $\vec{u}_k$ is 122. We also normalize the eigenvalue vector w.r.t. the total value of eigenvalues, i.e., $\overline{\boldsymbol{\Lambda}}_{1\times24} = \boldsymbol{\Lambda}_{1\times24}/\lambda_t$, where $\lambda_t = \lambda_1 + \lambda_2 + ... + \lambda_{24}$. The normalization for eigenvalues is required to ensure that they have the same scale as the F values. Thereafter, we use $\overline{F}, \overline{\Lambda}$ and U to build the feature importance matrix $\boldsymbol{I}_{122\times24} = [\overline{F}_1\overline{\lambda}_1\vec{u}_1, \overline{F}_2\overline{\lambda}_2\vec{u}_2, ..., \overline{F}_{24}\overline{\lambda}_{24}\vec{u}_{24}]$. In the end, we sum up the absolute value of each element in every row of the matrix $\boldsymbol{I}_{122\times24}$,

$$\vec{I}_{122\times1} = \sum_{n=1}^{24} |\overline{F}_n\overline{\lambda}_n\vec{u}_n|$$

$\vec{I}_{122\times1}$ is the feature importance vector in the original data space, and we also normalize it w.r.t. its total importance and rescale it,

$$\vec{I}_{nrm} = 100 \cdot \vec{I}_{122\times1} / \sum_{m=1}^{122} \vec{I}_m$$

Now, $\vec{I}_{nrm}$ is the normalized feature importance vector in the original data space of dimension 122, and each value of this vector represents the importance of a brain tissue volume in a region for the task of predicting the Gf scores. Higher importance represents higher correlation to the Gf score.

3 Results and Discussion

We examine the Gf prediction performance of individual models and StackNet on the combined dataset with 10-fold cross-validation, with the quantitative results shown in Table 1. The baseline is calculated by assigning the mean fluid intelligence ($\mu = 0$) to every subject in the combined dataset. From Table 1, the performance of each model is better than the baseline of guessing the mean every subject, and the performance of the StackNet is better than every individual model within itself because it takes advantage of stacked generalization. The proposed StackNet achieves a MSE of 94.2525 on the testing data as reported in the final leader board.

Table 1. The quantitative results of 11 models and StackNet with 10-fold cross-validation on the combined dataset. The bold number highlights the best performance.

Model	MSE
Baseline	84.50
BayesianRidge	82.62
Ridge(alpha = 512)	82.61
KernelRidge(alpha = 512)	82.61
GradientBoostingRegressor(n_estimators = 40, max_depth = 3)	83.60
RandomForestRegressor(n_estimators = 1000, max_depth = 7)	83.07
RandomForestRegressor(n_estimators = 1000, max_depth = 9)	83.09
RandomForestRegressor(n_estimators = 800, max_depth = 11)	83.07
RandomForestRegressor(n_estimators = 800, max_depth = 13)	83.11
ExtraTreesRegressor(n_estimators = 1800, max_depth = 9)	83.16
ExtraTreesRegressor(n_estimators = 2200, max_depth = 11)	83.10
ExtraTreesRegressor(n_estimators = 3200, max_depth = 15)	83.16
StackNet	**82.42**

The proposed StackNet in Fig. 1 is different from the StackNet which is used to report the MSE on the validation leader board. The StackNet used to report the MSE on the validation leader board has two layers and 8 models, and it achieves an MSE of 84.04 and 70.56 (rank 7 out of 17 teams) on the training and validation set, respectively. However, we noticed that statistics between the training set and validation are quite different, so we decided to combine these two datasets and work on this combined dataset ($\mu = 0$ and $\sigma = 9.19$) using 10-fold cross-validation. In addition, we also ensured that the mean and standard deviation of each fold is similar to the mean and standard deviation of combined dataset.

Second, we compute the importance of each dimension of the extracted feature by leveraging the F score from feature selection and eigenvectors and eigenvalues from PCA as described in Sect. 2.4. Each dimension of the extracted feature corresponds to the volume of a certain type of brain tissue in a certain region. Tables 2 and 3 show the top 10 most and least important feature dimensions for the task of predicting Gf, respectively, and higher importance represents higher correlation to the Gf scores.

Deep Learning vs. Classical Machine Learning: A Comparison of Methods for Fluid Intelligence Prediction

Luke Guerdan[(✉)], Peng Sun, Connor Rowland, Logan Harrison,
Zhicheng Tang, Nickolas Wergeles, and Yi Shang

University of Missouri - Columbia, Columbia, MO 65021, USA
{lmg4n8,ps793,carfzf,lkh6yb,zt253,wergelesn,shangy}@missouri.edu

Abstract. Predicting fluid intelligence based on T1-weighted magnetic resonance imaging (MRI) scans poses several challenges, including developing an adequate data representation of three dimensional voxel data, extracting predictive information from this data representation, and devising a model that is able to leverage the predictive information. We evaluate two strategies for prediction of fluid intelligence given structural MRI scans acquired through the Adolescent Brain Cognitive Development (ABCD) Study: deep learning models trained on raw imagery and classical machine learning models trained on extracted features. Our best-performing solution consists of a classical machine learning model trained on a combination of provided brain volume estimates and extracted features. Specifically, a Gradient Boosting Regressor (GBR) trained on a PCA-reduced feature space produced the best performance (train MSE = 66.29, validation MSE = 70.16), surpassing regression models trained on the provided volume data alone, and 2D/3D Convolutional Neural Networks trained on various representations of imagery data. Nonetheless, these results remain slightly better than a mean prediction, suggesting that neither approach is capturing a high degree of variance in the data.

Keywords: Image processing · Neurocognitive prediction · Machine learning

1 Introduction

An ongoing challenge in neuroscience is relating brain structure to function, both at a neural scale and at the level of phenotypic expression. Though new neuroimaging approaches such as functional magnetic resonance imaging (fMRI) and diffusion tensor tractography have begun to shed light into this area, relating basic structural properties to complex behavioral expression remains difficult [1–3]. T1-weighted MRI is one neuroimaging method which has been used to

---

L. Guerdan and P. Sun—denotes equal contribution.

© Springer Nature Switzerland AG 2019
K. M. Pohl et al. (Eds.): ABCD-NP 2019, LNCS 11791, pp. 17–25, 2019.
https://doi.org/10.1007/978-3-030-31901-4_3

relate brain structure to progression of autism, Alzheimer's, and Parkinson's [4–6]. However, these abnormalities are often associated with gross differences in brain tissue, and the work employing structural MRI to discern more nuanced differences in regular brain function remains limited. Methodological advances in this area could prompt neuroscientific discoveries and objective assessment methods for normal and abnormal brain function.

One specific behavioral measure presumably tied to brain structure is fluid intelligence, especially fluid intelligence, which is central in abstract reasoning and problem solving [7]. Fluid intelligence has been linked to fronto-parietal connection properties [9], and a number of other brain characteristics [10]. Yet, no literature found by the authors directly links MRI-based structural information to fluid intelligence. A large corpus of data available through the Adolescent Brain Cognitive Development (ABCD) Study provides an unprecedented opportunity to investigate potential links between fluid intelligence and structural features of the brain. What's more, the ABCD Neurocognitive Prediction Challenge offers a structured context in which to develop pipelines for inferring fluid intelligence scores from scans acquired in the ABCD study. This work proposes a prediction pipeline addressing this challenge consisting of a GBR trained on ROI shape features. Results based on other regressors trained on extracted shape features, and results based on CNNs trained on representations of raw imagery, are also provided for comparison.

2   Related Work

A key issue to be addressed in a neurocognitive prediction context is the information representation, as MRI data natively occupies a three dimensional voxel space. One basic approach is to consider volume estimates (or other derived features) of brain regions, as the relative size and form of each region may contribute to cognitive function. Yet another approach is to examine patterns in the raw MRI imagery using computer vision methods. The information representation able to best capture variance in the data related to fluid intelligence may lead to the best results.

2.1   Raw Imagery

A recent trend in computer vision is replacing hand-engineered features with the unprocessed image to create an end-to-end prediction pipeline. This approach has proved particularly effective when combined with deep learning methods such as convolutional neural networks (CNNs), which have demonstrated the ability to extract salient features directly from the input [13]. However, most of these methods have been applied in the context of a two dimensional input image, as opposed to a 3D voxel space. One approach is to slice into the three dimensional space, however, this raises the issue of which slice is most discriminative, and may leave valuable information un-processed. A 2.5D CNN, which ingests slices from the axial, coronal, and sagittal planes, is one solution addressing this problem. This approach has proven useful for generating state-of-the-art brain segmentation results [16]. Yet, this strategy also poses the challenge of the optimal slice to extract from each dimension. This motivates a 3D CNN which can ingest full imagery data from the entire voxel space. A 3D CNN has been employed for Alzheimer's disease diagnostics, and demonstrated high predictive accuracy [5].

2.2   Derived Features

One direct method of representing brain structure is calculating features based on brain regions of interest, then training a machine learning model on these features [4, 8, 12]. For example, a deep autoencoder trained on brain region volume features predicted autism onset with high accuracy [4]. Another work successfully leveraged global and local shape features of tumor areas to predict benign vs. malignant tumor regrowth [12]. Specifically, this approach extracted global features such as elongation, compactness, volume, and surface area, as well as local characteristics calculated from Gaussian and mean curvatures of a constructed isosurface. Global shape features were calculated using the Insight Segmentation and Registration Toolkit (ITK) implementation [14]. Not only did this method achieve high accuracy, but it also conferred the advantage of revealing which features are most discriminative of tumor malignancy [12]. The interpretability of deep learning methods, especially those leveraging raw imagery data, is less direct.

3   Methods

Approaches for predicting fluid intelligence based on MRI imagery based on both (a) raw imagery, and (b) derived features were developed and tested. We predicted that deep learning methods ingesting raw MRI imagery would ultimately have more representational power and would demonstrate better results. However, due to the complexity of the raw MRI imagery, the deep learning models we tested did not yield better results with this dataset. Though classical machine learning models trained on derived features provided lower mean squared error, we still provide details regarding the deep learning methods developed for comparison.

3.1   Deep Learning Methods

In order to extract shape information for each brain region, we convolved across voxels in three dimensions using a 3D CNN. This is motivated by recent work indicating CNNs can extract meaningful shape patterns in image classification [15] and object detection [11] contexts. Our pipeline for deep learning based on raw imagery is shown in Fig. 1. We adapted a 3D-ResNet [18], which was originally used for video classification, to integrate information across slices.

Two specific adaptions were required. (1) Since the third dimension is temporal correlation in a video classification task, this was changed to spatial correlation over the input shape. (2) The final layer was updated to consist of a single node outputting the predicted fluid intelligence score.

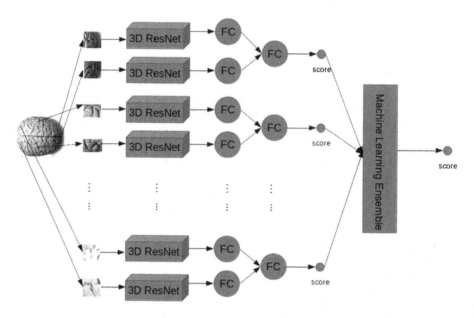

Fig. 1. 3D CNN for prediction of fluid intelligence score for each brain region. Each hemisphere is fed into a 3D ResNet separately and then results are stacked into one connected layer together. The score predicted from each region (both left and right hemispheres) is then aggregated using a classical machine learning ensemble method to generate a final score.

As shown in Fig. 1, feature extraction from each provided region of segmented grey matter [17] was first conducted using the previously described 3D CNN. The proceeding fully connected layers were organized hierarchically, with the first layer matching extracted features across hemispheres. A model was trained for each brain region (including left and right hemisphere), then predicted scores were ensembled using a regression model. A preliminary assessment of regression models including AdaBoost, Ridge regression, Random Forest, and SVR showed that AdaBoost provided the best performance, which was used in the reported experiments. This was performed for both grey matter and non-grey matter in the region of interest.

3.2 Shape-Based Machine Learning Methods

We also developed a preprocessing pipeline which ingests the skull stripped and raw MRI images provided for each participant, and extracts global shape features based on isolated ROIs, full grey matter, and full non-grey matter as calculated by [17]. Version 4.13.1 of ITK was used to extract (a) volume, (b) elongation, (c) surface area, (d) roundness, and (e) flatness from each of the 100 largest regions provided in the segmented grey matter [14].

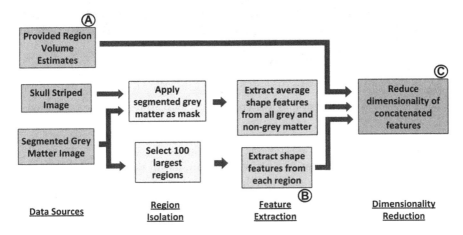

Fig. 2. Pipeline for extracting and aggregating features used in regressor training (classical machine learning models). **A.** Shows the dataset for provided estimates alone, **B.** Shows the dataset consisting of only ROIs, while **C.** Shows the dataset consisting of all features.

The ordering of these 100 regions was fixed based on the first training participant and kept constant for the training, validation, and test data. ITK returns extracted features for each of the non-contiguous segments of the isolated ROI, and we used the first (largest) segment returned. Shape features from the overall grey and non-grey matter were extracted according to the same process above, except that each shape feature was averaged for all segments returned. Non-grey matter was isolated by applying the grey matter as a mask to the overall skull-stripped image for each participant (shown in Fig. 3). This non-grey matter was included to provide some proxy of white matter, which has been shown to be a relevant factor for fluid intelligence in the literature. Including this information empirically showed a slight performance improvement in validation loss.

The feature extraction process resulted in a 500 dimensional feature space generated by the ROI shape extraction, and a 10 dimensional feature space generated by global feature extraction. Each of these features were then combined with the provided volume estimates to give a final dataset for training. Since this dataset occupied a high-dimensional space, dimensionality reduction (by PCA) was used to project the data to a lower dimensional subspace. A series of regression models were then trained on the PCA-reduced final dataset (Fig. 2C), as well as the given volume data alone (Fig. 2A), and the extracted ROI features alone (Fig. 2B). The specific models trained included lasso, ridge, support vector, gradient boosting, and AdaBoost regressors. A grid search was performed

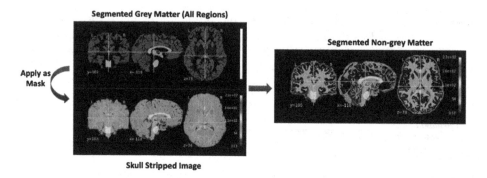

Fig. 3. Process of isolating grey and non-grey matter for feature extraction using masking.

to determine the optimal parameters for training on all datasets, and PCA was included in this parameter grid search for the (Fig. 2C) dataset. Models were evaluated by the MSE error metric, which defines the error between the predicted labels $\hat{Y}$ and actual labels Y over n samples as:

$$MSE = \frac{1}{n}\sum_{i=1}^{n}(Y_i - \hat{Y}_i)^2 \tag{1}$$

The train, validation, test split aligned with the data provided in the ABCD Neurocognitive Prediction Challenge, with 3739 samples used for training, 415 samples used for validation, and 4402 samples left out for testing. In order to establish a baseline estimate of the predictive value of each model, the error of predicting the mean was first established by calculating the average score from the training dataset and calculating the MSE between this mean and each label in the training and validation datasets. A series of regressors were then trained on each of the three datasets described above, including Ridge and Lasso regression, SVR, GBR, and AdaBoost.

4   Results

Table 1 reports training and validation error obtained from training the five algorithms above on dataset A, B, and C (Fig. 2), as well as results from the best-performing 3D CNN models. Figure 4 shows a visual comparison of validation loss provided in the table. Overall, the GBR demonstrated the highest predictive value for each of the three extracted feature datasets, with the best performing model being a GBR trained on the PCA-reduced full features. This reached a $MSE(train) = 66.29$ and $MSE(test) = 70.16$. The optimal configuration for this model consisted of 95 PCA components and 45 learners in the GBT. Compared with 3D CNN models, as shown in Table 1, the GBT outperformed 3D CNN models by 1.14 MSE in validation error.

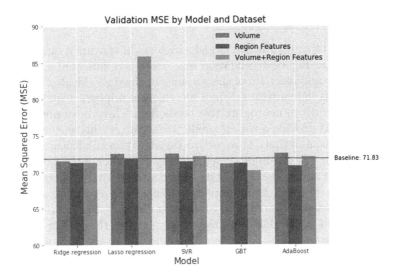

Fig. 4. Validation MSE for each of the five models trained on the three datasets. The horizontal bar indicates the MSE obtained by predicting the mean of the training dataset.

Table 1. Train and validation MSE obtained from all methods and datasets.

Dataset	Method	Train (MSE)	Val. (MSE)
	Mean prediction	**85.84**	**71.83**
Volume (A)	Ridge reg.	82.64	71.52
	Lasso reg.	81.62	72.53
	SVR	83.06	72.51
	GBR	80.16	71.1
	AdaBoost	79.60	72.49
Extracted features (B)	Ridge reg.	78.28	71.22
	Lasso reg.	85.84	71.83
	SVR	77.88	71.43
	GBR	73.50	71.21
	AdaBoost	77.85	70.79
Volume + Extracted features (C)	Ridge reg.	81.84	71.22
	Lasso reg.	71.83	85.84
	SVR	72.22	72.16
	GBR	**66.29**	**70.16**
	AdaBoost	80.10	72.03
White matter imagery	3D CNN	85.43	71.64
Grey matter imagery	3D CNN	85.52	71.75
White+Grey model imagery	3D CNN ensemble	84.55	71.30

5   Discussion

Overall, it proved difficult to find a model which could capture a large degree of variance in the data. The best validation score obtained indicated a R2 value of .03, which captures a modest 3% of the variance in the data. Most of the derived features and models demonstrate menial incremental improvements as opposed to qualitative advances in neurocognitive prediction. Though we predicted that deep learning would provide a robust predictive strategy, this was not the case. We suspect the reason is that (1) excessive variability in the features extracted between subjects or (2) the input images are too complex for a CNN to learn relevant features. Though the ABCD dataset contains a large amount of data for a neuroscience context, this quantity still falls very short of the enormous corpuses typically used in other image processing domains. Therefore, additional data could help aid prediction. However, it is possible that given the correct way of representing the data, better results could be obtained. Models such as autoencoder could also provide predictive value, as they have in similar contexts [4].

It is also possible that structural features alone do not contain enough information related to fluid intelligence to be useful in prediction contexts. Other information such as cortical connectivity and functional activation may be required to capture an accurate assessment of ones fluid intelligence. Combining multi-modal neuroimaging datasets presents yet another representational problem, but could prove fruitful in bolstering model performance.

References

1. Le Bihan, D., et al.: Diffusion tensor imaging: concepts and applications. J. Magn. Reson. Imaging Off. J. Int. Soc. Magn. Reson. Med. **13**(4), 534–546 (2001)
2. Raichle, M.E., MacLeod, A.M., Snyder, A.Z., Powers, W.J., Gusnard, D.A., Shulman, G.L.: A default mode of brain function. Proc. Nat. Acad. Sci. **98**(2), 676–682 (2001)
3. Bullmore, E., Sporns, O.: Complex brain networks: graph theoretical analysis of structural and functional systems. Nat. Rev. Neurosci. **10**(3), 186 (2009)
4. Hazlett, H.C., et al.: Early brain development in infants at high risk for autism spectrum disorder. Nature **542**(7641), 348 (2017)
5. Hosseini-Asl, E., Keynton, R., El-Baz, A.: Alzheimer's disease diagnostics by adaptation of 3D convolutional network. In: 2016 IEEE International Conference on Image Processing (ICIP), pp. 126–130. IEEE, September 2016
6. Morales, D.A., et al.: Predicting dementia development in Parkinson's disease using Bayesian network classifiers. Psychiatry Res. NeuroImaging **213**(2), 92–98 (2013)
7. Stankov, L.: Complexity, metacognition, and fluid intelligence. Intelligence **28**(2), 121–143 (2000)
8. Sun, P., et al.: Ada-automatic detection of alcohol usage for mobile ambulatory assessment. In: 2016 IEEE International Conference on Smart Computing (SMARTCOMP). IEEE (2016)
9. Lee, K.H., et al.: Neural correlates of superior intelligence: stronger recruitment of posterior parietal cortex. Neuroimage **29**(2), 578–586 (2006)

10. Haier, R.J., Jung, R.E., Yeo, R.A., Head, K., Alkire, M.T.: Structural brain variation and general intelligence. Neuroimage **23**(1), 425–433 (2004)
11. Liu, Y., et al.: Performance comparison of deep learning techniques for recognizing birds in aerial images. In: 2018 IEEE Third International Conference on Data Science in Cyberspace (DSC). IEEE (2018)
12. Ismail, M., et al.: Shape features of the lesion habitat to differentiate brain tumor progression from pseudoprogression on routine multiparametric MRI: a multisite study. Am. J. Neuroradiol. **39**(12), 2187–2193 (2018)
13. Krizhevsky, A., Sutskever, I., Hinton, G.E.: ImageNet classification with deep convolutional neural networks. In: Advances in Neural Information Processing Systems, pp. 1097–1105 (2012)
14. Kim, Y.: Insight segmentation and registration toolkit. The National Library of Medicine, Washington, DC (2001)
15. Chen, G., Sun, P., Shang, Y.: Automatic fish classification system using deep learning. In: 2017 IEEE 29th International Conference on Tools with Artificial Intelligence (ICTAI), pp. 24–29. IEEE, November 2017
16. Kushibar, K., et al.: Automated sub-cortical brain structure segmentation combining spatial and deep convolutional features. Med. Image Anal. **48**, 177–186 (2018)
17. Pfefferbaum, A., et al.: Altered brain developmental trajectories in adolescents after initiating drinking. Am. J. Psychiatry **175**(4), 370–380 (2017)
18. Hara, K., Kataoka, H., Satoh, Y.: Can spatiotemporal 3D CNNs retrace the history of 2D CNNs and imagenet? In: Proceedings of the IEEE Conference on Computer Vision and Pattern Recognition (2018)

Surface-Based Brain Morphometry for the Prediction of Fluid Intelligence in the Neurocognitive Prediction Challenge 2019

Michael Rebsamen[1,2]([✉])(iD), Christian Rummel[1], Ines Mürner-Lavanchy[3](iD), Mauricio Reyes[4](iD), Roland Wiest[1], and Richard McKinley[1](iD)

[1] Support Center for Advanced Neuroimaging (SCAN),
University Institute of Diagnostic and Interventional Neuroradiology,
University of Bern, Inselspital, Bern University Hospital, Bern, Switzerland
michael.rebsamen@insel.ch
[2] Graduate School for Cellular and Biomedical Sciences,
University of Bern, Bern, Switzerland
[3] University Hospital of Child and Adolescent Psychiatry and Psychotherapy,
University of Bern, Bern, Switzerland
[4] Healthcare Imaging A.I. Lab, Insel Data Science Center,
Inselspital, Bern University Hospital, Bern, Switzerland

Abstract. Brain morphometry derived from structural magnetic resonance imaging is a widely used quantitative biomarker in neuroimaging studies. In this paper, we investigate its usefulness for the Neurocognitive Prediction Challenge 2019.

An in-depth analysis of the features provided by the challenge (anatomical segmentation and volumes for regions of interest according to the SRI24 atlas) motivated us to process the native T1-weighted images with FreeSurfer 6.0, to derive reliable brain morphometry including surface based metrics. A combination of subcortical volumes and cortical thicknesses, curvatures, and surface areas was used as features for a support-vector regressor (SVR) to predict pre-residualized fluid intelligence scores. Results performing only slightly better than the baseline (uniformly predicting the mean) were observed on two internally held-out validation sets, while performance on the official validation set was approximately the same as the baseline.

Despite a large dataset of a specific cohort available for training, this suggests that structural brain morphometry alone has limited power for this challenge, at least with today's imaging and post-processing methods.

Keywords: Brain morphometry · Structural MRI · Neurocognition · Machine learning

K. M. Pohl et al. (Eds.): ABCD-NP 2019, LNCS 11791, pp. 26–34, 2019.
https://doi.org/10.1007/978-3-030-31901-4_4

1   Introduction

The Adolescent Brain Cognitive Development (ABCD) study aims to investigate factors influencing the development of the brain and cognition in adolescence. The ABCD Neurocognitive Prediction Challenge (ABCD-NP-Challenge 2019) leverages data from ABCD and asks participating teams to predict fluid intelligence [1] from magnetic resonance imaging (MRI). Although functional MRI data is also available for the cohort, the challenge is formulated explicitly to predict pre-residualized fluid intelligence scores solely based on structural MRI, namely T1-weighted (T1w) images.

Studies have reported correlations of fluid intelligence with cortical gray matter volume [16,18,19], cortical thickness [21,23], surface area [8], and the shapes of subcortical structures in the basal ganglia [5]. Rates of changes in the cortex due to reorganization and synaptic pruning are thought to be an indicator for plasticity of the brain in adolescence and are associated with cognitive development [28].

Brain morphometry, which is widely used to analyze brain development [15], requires accurate and unbiased reconstruction methods. Given the suggested relevance of surface-based morphometry in the literature, and our analysis of the provided atlas-based segmentation data, we decided to post-process the raw MRI with FreeSurfer (FS) 6.0 [12]. In addition to describing the final method used in the challenge, we explicitly mention some unsuccessful attempts to predict the scores both from surface-based morphometry and directly from the images. We also present a comparison of metrics derived from FreeSurfer with the features provided by the challenge.

2   Materials and Methods

Data were provided by the ABCD-NP-Challenge and stem from 9–10 year old children recruited across the United States. The MR images were acquired on three different 3T-scanners across 21 data collection sites with an isotropic 1 mm resolution [6]. Information from post-processing the images were provided by the challenge additionally to the raw images. Namely segmentations parcellated into regions of interest according to the SRI24 atlas [27] along with the corresponding volumes. After skull-stripping, tissue segmentations were generated using Atropos [3] and registered to the atlas with ANTS [2] (see data supplement of Pfefferbaum et al. [25]).

The fluid intelligence scores were pre-residualized on age, gender, ethnicity, parental education, parental marital status, data collection site, and brain volume. A total of 3739 samples were available for training and 415 for validation, whereas the scores for the 4402 samples in the test set had to be predicted by the participating teams, who will be ranked by mean squared error (MSE).

2.1   Surface-Based Morphometry

We processed the raw T1w MRI with FreeSurfer 6.0 (`recon-all`, including hippocampal subsegmentation). Without manual interventions, the failure rate was

0.6% (primarily failures correcting topological errors during surface reconstruction) and required an average processing time of 13.0 ± 2.8 hours per subject. For subjects where more than one scan was available, the image with the most recent time stamp was used.

FreeSurfer segmentations from assigning each voxel one of 37 neuroanatomical labels [13] yield volumetric information for tissue classes including ventricles and subcortical gray-white matter structures. Metrics derived from surface-based morphometry [9] include mean and standard deviation of cortical thickness, mean and Gaussian curvature, curvature and folding index, surface area, and gray matter volume on a per voxel level. We have used their averages per region of interest (ROI) as defined by the atlases of Desikan-Killiany (DK) [10] and Destrieux [11] parcellated into 34 and 74 ROIs per hemisphere, respectively.

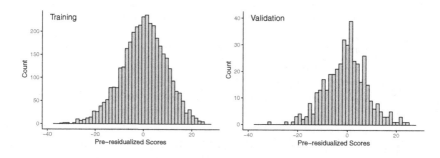

Fig. 1. Distribution of ground truth in the training and validation set.

2.2   Analysis of Provided Scores

The distributions of the scores in Fig. 1 show significant left skewness (-0.24) in the training set ($p < 10^{-8}$ of a D'Agostino skewness test) which is not present in the validation set ($p = 0.768$). An F-test suggests a significant difference in the variances ($F = 0.835, p = 0.017$) which we interpret as an indication that the validation set is not a random sample from the same distribution as the training set. This finding motivated us to carve out two additional internal validation sets (N $= 464$), randomly sampled from the training data.

We analyzed the provided segmentations and compared them to the results from FreeSurfer (Fig. 2). Of particular interest is the supratentorial volume which is used as *brain volume* by the challenge organizers in building the linear model to calculate the pre-residualized scores. The supratentorial volume from Atropos/SRI24 is only weakly but significantly correlated to the corresponding volume from FS (Pearson's $r = 0.149$, $p < 10^{-15}$) and has much smaller inter-subject variance (Fig. 2d). Although it is reasonable to assume that brain volume increases in preadolescence [14,15], the supratentorial volume from Atropos is negatively correlated with age ($r = -0.038, p = 0.022$) whereas the FreeSurfer

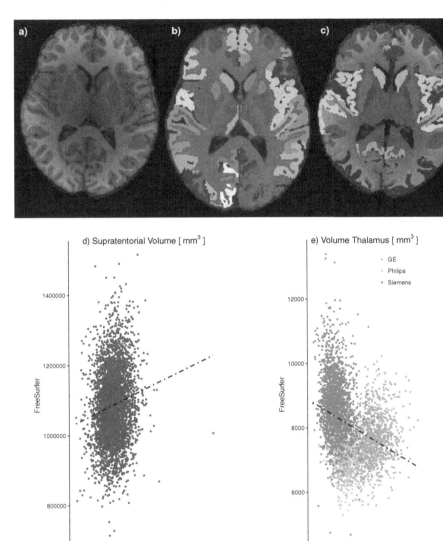

Fig. 2. First row: example segmentation (subject NDAR_INV05WK8AN7) as provided by the challenge from Atropos normalized to SR24 (middle) and from FreeSurfer (right). Undersegmentation of thalamus, putamen and superior occipital gyrus, and oversegmentation of superior frontal gyrus by Atropos was observed. Last row: correlation of volumes between challenge provided segmentation and FreeSurfer for the supratentorial volume (left) and right thalamus (right). Colors representing scanner. (Color figure online)

equivalent is positively correlated ($r = 0.035, p = 0.034$). A more reliable estimator for total brain volume is FreeSurfer's estimated total intracranial volume (eTIV) [4] which is more strongly correlated with age ($r = 0.090, p < 10^{-7}$). A negative correlation between Atropos and FS was observed for the volume of the thalamus, a structure whose derived segmentation is obviously biased by the scanner (Fig. 2e).

Table 1 compares two linear models to calculate the pre-residualized scores, one with supratentorial volume from Atropos as brain volume and another with eTIV from FS. Although the coefficient for volume becomes positive with eTIV, the resulting scores of the two models still show high agreement ($R^2 = 0.997$).

Table 1. Coefficients of a linear model as described by the challenge, with supratentorial volume from Atropos (left) and with FreeSurfer's estimated total intracranial volume (eTIV) as brain volume. Further 19 site factors which were not significant are omitted for space reasons. Signif. codes: $*** < 0.001 < ** < 0.01 < * < 0.05 < . < 0.1$

	LM Suptent.		LM with FS eTIV	
	Estimate	p	Estimate	p
(Intercept)	44.222	<2e−16 ***	44.428	<2e−16 ***
age	0.361	<2e−16 ***	0.358	<2e−16 ***
femaleyes	1.204	3.41e−05 ***	1.672	1.87e−07 ***
race.ethnicityHispanic	−0.227	0.633	−0.112	0.813
race.ethnicityBlack	−4.740	<2e−16 ***	−4.462	1.45e−14 ***
race.ethnicityAsian	3.041	0.002 **	3.269	0.001 **
race.ethnicityOther	−0.278	0.588	−0.201	0.695
high.educHS Diploma/GED	1.824	0.050 .	1.831	0.049 *
high.educSome College	3.582	1.25e−05 ***	3.569	1.31e−05 ***
high.educBachelor	5.919	2.14e−12 ***	5.837	4.09e−12 ***
high.educPostGrad. Deg.	6.724	1.41e−15 ***	6.590	4.91e−15 ***
marriedyes	0.632	0.075 .	0.582	0.100
abcd_sitesite08	2.459	0.051 .	2.876	0.022 *
abcd_sitesite15	−3.159	0.011 *	−3.384	0.007 **
(non sign. sites omitted)				
volume	−0.211	0.159	0.619	7.95e−4 ***

Statistical analyses were performed using R [26] with the *stats* package version 3.5.1. A significance level of $\alpha = 0.05$ was used unless otherwise mentioned. The metric R^2 refers to coefficient of determination (and not Pearson's r squared as used in the challenge).

2.3   Prediction of Fluid Intelligence Scores

From a statistical analysis of all available cortical morphometric parameters from FreeSurfer, 9% correlate with the pre-residualized score. After Bonferroni correction for multiple comparisons, this reduced to three parameters: Gaussian curvature ($r = -0.08, p = 1.3 \times 10^{-6}$) and curvature index ($r = -0.07, p = 9.1 \times 10^{-6}$) in the left middle temporal gyrus and Gaussian curvature in the right superior temporal gyrus ($r = -0.07, p = 1.4 \times 10^{-5}$).

We initially tried an approach based on classification: if one could reliably identify subjects belonging to the upper or lower quartile, and predict their scores with the mean of the corresponding quartile (predicting zero for the remaining subjects), a MSE of 14.4 would be reached on the validation set. However, an attempt to classify just the 5% and 95% quantiles with a support vector machine (SVM) reached an accuracy of only 60%.

We trained support vector regressors (SVR) for the prediction of the scores on five feature sets (Table 2), each based on combinations of subcortical volumes and cortical measures from two different atlases, using Python 3.6 and scikit-learn [24]. The features were centered to zero mean and scaled to unit variance. An *epsilon-SVR* [7] with a radial basis function (RBF) kernel was chosen. Hyperparameters were determined with a 4-fold cross validation on the internal training set, yielding a regularization term $C = 1.6$ and model capacity $\gamma = $ auto. The final models were trained on the internal training set with 2777 samples.

Additionally, we trained a 3D convolutional neural network (CNN) to directly predict the pre-residualized fluid intelligence scores from the raw, skull-stripped T1 images. The network consisted of three convolutional layers and three fully connected layers. For the final submission, we selected the method with the best average mean squared error (MSE) over all three validation sets.

3   Results

A baseline of predicting the mean of the ground truth scores of the training set leads to an MSE of 71.640 on the validation set. The predictions of an SVR trained on each of the five feature sets performed slightly better than this baseline on the two internally held-out validation sets (Table 2). However, only feature set F5, using all available metrics from FreeSurfer, performed better on the official validation set. The best average MSE over all three sets was reached using F3, which was therefore used for the final submission. This model reached an MSE of 93.0326 on the official test set, which is slightly better than a naive baseline predictor (94.695).

4   Discussion

We have used FreeSurfer to generate anatomical segmentations and surface-based morphometric measures from raw T1w images, a method that is known to produce results with high reliability [20, 22]. Despite a large amount of data available for training machine-learning approaches with these brain morphometry features, results for predicting pre-residualized fluid intelligence scores were only marginally better than the baseline. Contrasting the observed performance

Table 2. Result overview for the various models evaluated on the three validation sets. Feature sets for SVR are F1: Volumes from challenge. F2: FreeSurfer volumes and cortical thickness, curvature and surface area from DK atlas. F3: same as F2 but with Destrieux atlas. F4: DK and Destrieux combined. F5: All available metrics as per methods section. Bold numbers indicate the best mean squared error (MSE) per set.

	Int. validation 1		Int. validation 2		Validation (challenge)	
	R^2	MSE	R^2	MSE	R^2	MSE
Baseline (predict mean)	−0.002	81.310	−0.000	81.437	−0.004	71.640
SVR F1	0.027	**78.946**	0.011	80.561	−0.030	73.699
SVR F2	0.021	79.498	0.008	80.806	−0.014	72.333
SVR F3	0.027	78.963	0.015	**80.235**	−0.005	71.706
SVR F4	0.027	78.968	0.013	80.391	−0.011	72.092
SVR F5	0.023	79.296	0.009	80.732	−0.003	**71.556**
3D CNN	0.010	80.335	0.012	80.479	−0.010	72.046

to a theoretical MSE of 14.4 that would be reached for a simplified, reliable classification of the tails of the distribution suggests a limited power of structural MRI for this type of challenge.

Significant inconsistencies in the volumes measured by the two different segmentation methods were observed, with some measures even contradicting each other, or showing inverse correlations to expected age trajectories. Structures like the thalamus are obviously biased by the scanner used, a factor that is known to influence brain morphometry [17]. Although MRI acquisition protocols have been optimized by the ABCD study to harmonize across the three scanners [6], the influence of scanner vendor on the segmentation methods may be worthwhile to investigate further.

Limitations: No systematic manual quality control of the results derived from FreeSurfer was performed, nor have we conducted manual corrections, potentially leading to outliers in the measures. Creation of a cohort-specific atlas could further improve the morphometry.

Outlook: The ABCD cohort will be expanded with longitudinal data, potentially leading to more robust predictions in the future: individual changes in morphometry may be correlated more strongly to fluid intelligence than cross-sectional measures [28]. We encourage the challenge organizers to review the provided ground truth, in particular the supratentorial volume used as brain size in the linear model.

Acknowledgements. Calculations were performed on UBELIX, the HPC cluster at the University of Bern.

References

1. Akshoomoff, N., et al.: VIII NIH toolbox cognition battery (CB): composite scores of crystallized, fluid, and overall cognition. Monogr. Soc. Res. Child Dev. **78**(4), 119–132 (2013). https://doi.org/10.1111/mono.12038
2. Avants, B.B., Tustison, N.J., Song, G., Cook, P.A., Klein, A., Gee, J.C.: A reproducible evaluation of ANTs similarity metric performance in brain image registration. Neuroimage **54**(3), 2033–2044 (2011). https://doi.org/10.1016/j.neuroimage.2010.09.025
3. Avants, B.B., Tustison, N.J., Wu, J., Cook, P.A., Gee, J.C.: An open source multivariate framework for n-tissue segmentation with evaluation on public data. Neuroinformatics **9**(4), 381–400 (2011). https://doi.org/10.1007/s12021-011-9109-y
4. Buckner, R.L., et al.: A unified approach for morphometric and functional data analysis in young, old, and demented adults using automated atlas-based head size normalization: reliability and validation against manual measurement of total intracranial volume. Neuroimage **23**(2), 724–738 (2004). https://doi.org/10.1016/j.neuroimage.2004.06.018
5. Burgaleta, M., et al.: Subcortical regional morphology correlates with fluid and spatial intelligence. Hum. Brain Mapp. **35**(5), 1957–1968 (2014). https://doi.org/10.1002/hbm.22305
6. Casey, B., et al.: The adolescent brain cognitive development (ABCD) study: imaging acquisition across 21 sites. Dev. Cogn. Neurosci. **32**, 43–54 (2018). https://doi.org/10.1016/j.dcn.2018.03.001
7. Chang, C.C., Lin, C.J.: LIBSVM: a library for support vector machines. ACM Trans. Intell. Syst. Technol. (TIST) **2**(3), 27 (2011). https://doi.org/10.1145/1961189.1961199
8. Colom, R., et al.: Neuroanatomic overlap between intelligence and cognitive factors: morphometry methods provide support for the key role of the frontal lobes. Neuroimage **72**, 143–152 (2013). https://doi.org/10.1016/j.neuroimage.2013.01.032
9. Dale, A.M., Fischl, B., Sereno, M.I.: Cortical surface-based analysis: I. Segmentation and surface reconstruction. Neuroimage **9**(2), 179–194 (1999). https://doi.org/10.1006/nimg.1998.0395
10. Desikan, R.S., et al.: An automated labeling system for subdividing the human cerebral cortex on MRI scans into gyral based regions of interest. Neuroimage **31**(3), 968–980 (2006). https://doi.org/10.1016/j.neuroimage.2006.01.021
11. Destrieux, C., Fischl, B., Dale, A., Halgren, E.: Automatic parcellation of human cortical gyri and sulci using standard anatomical nomenclature. Neuroimage **53**(1), 1–15 (2010). https://doi.org/10.1016/j.neuroimage.2010.06.010
12. Fischl, B.: FreeSurfer. Neuroimage **62**(2), 774–781 (2012). https://doi.org/10.1016/j.neuroimage.2012.01.021
13. Fischl, B., et al.: Whole brain segmentation: automated labeling of neuroanatomical structures in the human brain. Neuron **33**(3), 341–355 (2002). https://doi.org/10.1016/S0896-6273(02)00569-X
14. Giedd, J.N., et al.: Brain development during childhood and adolescence: a longitudinal MRI study. Nat. Neurosci. **2**(10), 861 (1999). https://doi.org/10.1038/13158
15. Giedd, J.N., Rapoport, J.L.: Structural MRI of pediatric brain development: what have we learned and where are we going? Neuron **67**(5), 728–734 (2010). https://doi.org/10.1016/j.neuron.2010.08.040

16. Haier, R.J., Jung, R.E., Yeo, R.A., Head, K., Alkire, M.T.: Structural brain variation and general intelligence. Neuroimage **23**(1), 425–433 (2004). https://doi.org/10.1016/j.neuroimage.2004.04.025

17. Han, X., et al.: Reliability of MRI-derived measurements of human cerebral cortical thickness: the effects of field strength, scanner upgrade and manufacturer. Neuroimage **32**(1), 180–194 (2006). https://doi.org/10.1016/j.neuroimage.2006.02.051

18. Kievit, R.A., et al.: Distinct aspects of frontal lobe structure mediate age-related differences in fluid intelligence and multitasking. Nature Commun. **5**, 5658 (2014). https://doi.org/10.1038/ncomms6658

19. Kievit, R.A., Fuhrmann, D., Borgeest, G.S., Simpson-Kent, I.L., Henson, R.N.: The neural determinants of age-related changes in fluid intelligence: a pre-registered, longitudinal analysis in UK Biobank. Wellcome Open Res. **3**, 38 (2018). https://doi.org/10.12688/wellcomeopenres.14241.2

20. Madan, C.R., Kensinger, E.A.: Test-retest reliability of brain morphology estimates. Brain Inform. **4**(2), 107–121 (2017). https://doi.org/10.1007/s40708-016-0060-4

21. Martínez, K., et al.: Reproducibility of brain-cognition relationships using three cortical surface-based protocols: an exhaustive analysis based on cortical thickness. Hum. Brain Mapp. **36**(8), 3227–3245 (2015). https://doi.org/10.1002/hbm.22843

22. Morey, R.A., Selgrade, E.S., Wagner, H.R., Huettel, S.A., Wang, L., McCarthy, G.: Scan-rescan reliability of subcortical brain volumes derived from automated segmentation. Hum. Brain Mapp. **31**(11), 1751–1762 (2010). https://doi.org/10.1002/hbm.20973

23. Naumczyk, P., et al.: Cognitive predictors of cortical thickness in healthy aging. In: Pokorski, M. (ed.) Clinical Medicine Research. AEMB, vol. 1116, pp. 51–62. Springer, Cham (2018). https://doi.org/10.1007/5584_2018_265

24. Pedregosa, F., et al.: Scikit-learn: machine learning in python. J. Mach. Learn. Res. **12**(Oct), 2825–2830 (2011)

25. Pfefferbaum, A., et al.: Altered brain developmental trajectories in adolescents after initiating drinking. Am. J. Psychiatry **175**(4), 370–380 (2017). https://doi.org/10.1176/appi.ajp.2017.17040469

26. R Core Team: R: a language and environment for statistical computing. R Foundation for Statistical Computing, Vienna, Austria (2018). https://www.R-project.org/

27. Rohlfing, T., Zahr, N.M., Sullivan, E.V., Pfefferbaum, A.: The SRI24 multichannel atlas of normal adult human brain structure. Hum. Brain Mapp. **31**(5), 798–819 (2010). https://doi.org/10.1002/hbm.20906

28. Shaw, P., et al.: Intellectual ability and cortical development in children and adolescents. Nature **440**(7084), 676 (2006). https://doi.org/10.1038/nature04513

Prediction of Fluid Intelligence from T1-Weighted Magnetic Resonance Images

Sebastian Pölsterl$^{(\boxtimes)}$, Benjamín Gutiérrez-Becker, Ignacio Sarasua, Abhijit Guha Roy, and Christian Wachinger

Artificial Intelligence in Medical Imaging (AI-Med),
Department of Child and Adolescent Psychiatry,
Ludwig Maximilian Universität, Munich, Germany
{sebastian,benjamin,ignacio,abhijit,christian}@ai-med.de

Abstract. We study predicting fluid intelligence of 9–10 year old children from T1-weighted magnetic resonance images. We extract volume and thickness measurements from MRI scans using FreeSurfer and the SRI24 atlas. We empirically compare two predictive models: (i) an ensemble of gradient boosted trees and (ii) a linear ridge regression model. For both, a Bayesian black-box optimizer for finding the best suitable prediction model is used. To systematically analyze feature importance our model, we employ results from game theory in the form of Shapley values. Our model with gradient boosting and FreeSurfer measures ranked third place among 24 submissions to the ABCD Neurocognitive Prediction Challenge. Our results on feature importance could be used to guide future research on the neurobiological mechanisms behind fluid intelligence in children.

1   Introduction

Fluid intelligence [3] is a neuroscientific concept that is closely related to, but distinct from, general intelligence. It refers to the ability to think logically and to solve novel problems. It is believed that fluid reasoning plays a central role in cognitive development from childhood to early adulthood and enables children to acquire other abilities [1]. Anatomically, it is widely believed that higher fluid intelligence is linked to the maturation of the prefrontal cortex [9,25]. Previous neuroscientific findings indicate that brain volumes in parietal, occipital, and temporal as well as frontal cortical areas are related to intelligence [10]. In particular, it has been demonstrated that a region in the anterior prefrontal cortex, known as the rostrolateral prefrontal cortex plays a role in fluid reasoning [5,25]. Moreover, cortical thickness in early childhood has been associated with increased intelligence [14,21].

The ABCD Neurocognitive Prediction Challenge 2019 focuses on studying the relationship between brain and behavioral measures by asking participants to infer fluid intelligence solely from structural T1-weighted magnetic resonance

© Springer Nature Switzerland AG 2019
K. M. Pohl et al. (Eds.): ABCD-NP 2019, LNCS 11791, pp. 35–46, 2019.
https://doi.org/10.1007/978-3-030-31901-4_5

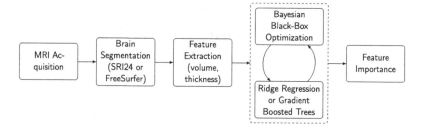

Fig. 1. Overview of our proposed pipeline for the prediction of fluid intelligence from T1-weighted MRI scans using SRI or FreeSurfer features.

images (MRI). To control for confounding factors, the organizers employed a residualized fluid intelligence score that accounts for a child's brain volume, age at baseline, sex at birth, ethnicity, highest parental education, parental income, parental marital status, and image acquisition site.

For extracting quantitative markers from MRI scans, we use two different approaches in this work. First, we use volumetric measurements provided by the challenge organizers based on the SRI24 atlas [18]. Second, we process MRI scans with FreeSurfer [6] for extracting volume and thickness measurements. In prior work, we obtained competitive results with FreeSurfer features for Alzheimer's prediction [2, 24]. Having access to two sets of features describing brain structure allows us to objectively evaluate which set provides more information for predicting fluid intelligence score. Moreover, we compare two approaches for regressing the fluid intelligence score: (i) a linear ridge regression, and (ii) gradient boosted trees. Linear ridge regression is a simple model, which offers the advantage of being less susceptible to overfitting. We use it as a baseline, because linear models are often used in clinical research due to being easy to interpret. Gradient boosted trees can model more complex relationships and are among the best performing prediction methods [26] and have been used in several winning entries to a wide range of Kaggle competitions.[1]

We employ data that was provided by The Adolescent Brain Cognitive Development(ABCD) Study [16], which recruited children aged 9–10. The residualized fluid intelligence scores and T1-weighted MRI scans of 3,736 children were provided to participants for training, and 415 samples for validation. Finally, data from 4,402 subjects were provided without fluid intelligence scores for testing. All data was obtained from the National Institute of Mental Health Data Archive.[2]

2   Methods

Figure 1 illustrates the components of our pipeline for the prediction of fluid intelligence from T1-weighted MRI scans. Scans were acquired according to the

---

[1] See https://github.com/dmlc/xgboost/blob/master/demo/README.md.

[2] https://nda.nih.gov/edit_collection.html?id=3104.

acquisition protocol of the Adolescent Brain Cognitive Development (ABCD) study protocol.[3] For quantifying brain morphometry, we used two different approaches: the SRI atlas and FreeSurfer. For SRI, we used volume measurements of 122 regions of interest extracted by the challenge's organizers based on the SRI24 atlas [18].[4] For FreeSurfer, we first performed automated segmentation of anatomical brain structures on MRI scans using FreeSurfer (version 5.3) [6]. We then extracted 136 volume and thickness measurements, which are all the measures produced by FreeSurfer scripts `asegstats2table` and `aparcstats2table`. These brain measures constituted the input to our machine learning models, linear ridge regression and gradient boosted trees [7,8]. For both, Bayesian black-box optimization for hyper-parameter tuning [22] was applied. Finally, we evaluated feature importance of the final models using the recently proposed SHapley Additive exPlanations (SHAP) [13].

2.1 Data-Preprocessing

We normalized all measurements while accounting for outliers by subtracting the median and dividing by the range between the 5% and 95% percentile. Thus, we reduce the impact of outliers and still obtain approximately centered features with equal scale. Finally, the provided residualized fluid intelligence scores in the training data where standardized to zero mean and unit variance; the same transformation as derived from the training data was applied to features and scores in the validation and test data.

2.2 Models

In following, we describe the two prediction models, where we used FreeSurfer features for the linear prediction, and FreeSurfer and SRI features for gradient boosting.

Linear Model. For the linear prediction of residualized fluid intelligence score, we combined features selected by automatic feature selection with features derived from literature. In particular, we included regions of the prefrontal cortex, because of results in previous studies [9,10,25]. For automatic feature selection, we employed univariate feature scoring by estimating the mutual information between each feature and the residualized fluid intelligence score [12]. First, we fitted a linear model based on FreeSurfer features describing the respective regions of interest in the prefrontal cortex. Next, we compared the estimated coefficients with the estimated feature importance by mutual information across all FreeSurfer features. After fusing this information, we ultimately selected eight features (see Table 3). Our final model was a ridge regression model [11], where we determined the strength of the ℓ_2 penalty by Bayesian black-box optimization

---

[3] https://abcdstudy.org/images/Protocol_Imaging_Sequences.pdf.

[4] See https://nda.nih.gov/data_structure.html?short_name=btsv01 for a full list of volumes.

(see next section). We fitted the model using the implementation in scikit-learn (version 0.18.2) [15].

Gradient Boosting. In addition, we selected stochastic gradient boosting [7,8] for predicting fluid intelligence. Gradient boosting performs functional gradient descent to find a function f that minimizes the squared loss:

$$\arg \min_f \quad \frac{1}{N} \sum_{i=1}^{N} (y_i - f(\mathbf{x}_i))^2, \tag{1}$$

where y_i denotes the standardized residualized fluid intelligence of the i-th subject and $\boldsymbol{x}_i$ is the feature vector of volume measurements derived from the MRI scan of the i-th subject. Gradient boosting constructs the function f in a greedy stagewise manner by fitting a simple base model g with parameters θ at the m-th iteration to the residuals of f from the previous iteration:

$$\arg \min_{\beta_m, \theta_m} \quad \frac{1}{N} \sum_{i=1}^{N} \left(y_i - f^{(m-1)}(\mathbf{x}_i) - \beta_m g(\mathbf{x}_i \,|\, \theta_m) \right)^2, \tag{2}$$

where $\beta_m \in \mathbb{R}$ is a weighting factor. The final estimated ensemble model after M iterations is given by

$$\hat{f}(\mathbf{x}) = \sum_{i=0}^{M} \beta_m g(\mathbf{x}_i \,|\, \theta_m). \tag{3}$$

Here, at each iteration, we employ a regression tree as base model g. To prevent overfitting, we fit the base model to a randomly selected subsample half the size of the whole training data. In addition, we use ℓ_1 regularization on the weights β_m to further reduce overfitting. We used the implementation available in XGBoost [4].

2.3   Hyper-parameter Selection

The performance of machine learning models depends to a large extent on the choice of hyper-parameters, e.g., the number of base learners in an ensemble. Traditionally, hyper-parameters have been selected by an expert or a grid search is performed. Grid search finds the best performing hyper-parameter configuration by evaluating all possible configurations selected a priori by an expert. However, it is often unknown a priori which values are suitable for a given machine learning task, which makes defining a list of values challenging. In contrast, Bayesian black-box optimization (see e.g. [22]) only requires specifying a prior distribution for each hyper-parameter. Most importantly, it can explore the whole domain of values and not only a list of pre-defined values. It has been shown that Bayesian hyper-parameter optimization outperforms grid search for

Table 1. Priors for hyper-parameters used in Bayesian hyper-parameter optimization. The first four rows refer to parameters from gradient boosting models, the last row is the only parameter optimized for the linear model. We denote a real-valued random variable drawn from a uniform distribution over the domain $[a, b] \subset \mathbb{R}$ as $X \sim \mathcal{U}(a, b)$. A random variable drawn from a log-uniform distribution is denoted as $X \sim \mathcal{LU}(a, b)$, where $X = \log_{10} Y$ with $Y \sim \mathcal{U}(10^a, 10^b)$.

Parameter	Prior
Iterations	$\mathcal{U}(10, 1000)$
Max. depth	$\mathcal{U}(1, 5)$
Learning rate	$\mathcal{LU}(10^{-5}, 1.25)$
ℓ_1 regularization	$\mathcal{LU}(10^{-6}, 2^{12})$
ℓ_2 regularization	$\mathcal{LU}(10^{-6}, 2^{12})$

many machine learning tasks [19]. We performed hyper-parameter search using scikit-optimize.[5]

For each proposed hyper-parameter configuration, we estimated the squared error over the validation set and used the sum of median and standard deviation of all errors as optimization criteria. We used Gaussian processes as surrogate models and probabilistically chose one of three acquisition functions at each iteration: (i) expected improvement (EI), (ii) lower confidence bound (LCB), or (iii) probability of improvement (PI). Initially, we assign each acquisition function equal weight $w_i = 0$ ($i = \{1, 2, 3\}$). For the t-th iteration, we independently optimized each acquisition function to propose a candidate hyper-parameter optimization $\mathbf{h}_i^{(t)}$. The final proposal $\mathbf{h}_*^{(t)}$ was selected probabilistically according to the softmax distribution with $p_i = \frac{w_i}{\sum_{k=1}^{3} w_i}$. After retrieving the corresponding prediction error and updating the surrogate model accordingly, we update the weights such that $w_i^{(t+1)} = w_i^{(t)} - \mathbb{E}(\mathbf{h}_*^{(t)})$. This process was repeated for 100 iterations. The best-performing hyper-parameter configuration was used for prediction. Our choice of prior distributions is summarized in Table 1.

2.4   Feature Importance

While gradient boosted trees are a potentially powerful model to solve a variety of prediction tasks, their black-box nature is often a barrier for the adoption of such model in the clinic. To be able to interpret complex non-linear machine learning methods, such as gradient boosting trees, we rely on Shapley values, which are a classic solution in game theory to determine the distribution of credits to players participating in a cooperative game [20, 23]. In particular, we employ the recently proposed SHAP (SHapley Additive exPlanations) values, which belong to the class of additive feature importance measures [13]. We describe the assignment of importance values to features in detail in [17].

---

[5] https://scikit-optimize.github.io.

3   Results

We describe the results for the three models: (i) linear model with FreeSurfer-based features, (ii) gradient boosting with SRI-based features, and (iii) gradient boosting with FreeSurfer-based features. The performance of all our models for the prediction of residualized fluid intelligence is summarized in Table 2. First, we are comparing models based on prediction performance. In the second part, we are inspecting models in more detail via SHAP values.

Table 2. Performance on training, validation and test set. MSE: mean squared error. MAE: mean absolute error. GBM: Gradient Boosting Model.

	Subjects	Linear		GBM (SRI24)		GBM (FreeSurfer)	
		MSE	MAE	MSE	MAE	MSE	MAE
Training	3,736	85.492	7.304	61.657	6.241	47.547	5.487
Validation	415	71.277	6.521	71.477	6.579	69.653	6.557
Test	4,402	93.215	—	94.103	—	92.563	—

3.1   Prediction Performance

Our linear model using FreeSurfer features ranked tenth among 24 submissions to the ABCD Neurocognitive Prediction Challenge with a difference of 1.0854 and 0.7179 to the first and second placed team, respectively. Results indicate that predicting residualized fluid intelligence from MRI-derived volume measurements is a challenging task for a linear model. In particular, the proposed model struggles to reliably predict residualized fluid intelligence at the extremes of the distribution, i.e., very low or very high values. Consequently, we observe a relatively high mean squared error, which is an order of magnitude larger than the mean absolute error. Interestingly, we obtained a lower error on the validation data than the training data. However, the error on the test data is relatively high. Although we do not have access to the test data, we believe the main increase in test error can be attributed to mispredictions at the extreme ends of the fluid intelligence distribution. We believe this to be a reasonable assumption since our model is a linear model with only few features, such that overfitting is unlikely to be a problem.

Our gradient boosting model based on SRI Features ranked 17th on the challenge's test data. Although gradient boosting allows for modelling non-linear relationships between MRI-derived features and residualized fluid intelligence, we note that Table 2 shows that the prediction error on the training set improved, but as before remains high, which indicates problems in predicting the extremes of the distribution. Moreover, worse performance on the validation and test set indicates that generalization to unseen data seems to be an issue to some extent. Unfortunately, we do not have access to the test data and cannot compute the mean absolute error. Therefore, it is unclear to which extent the relatively high mean squared error on the test set is due to overfitting on the training data or due to larger errors at the extreme ends of the fluid intelligence distribution.

By using gradient boosting with FreeSurfer features, we were able to achieve lower prediction error than with SRI-based features (see Table 2). Our model ranked third place among 24 submissions to the ABCD Neurocognitive Prediction Challenge with a difference of 0.0652 and 0.4327 to the second and first placed team, respectively. The model outperformed the linear regression model from above by 1.573 on the validation data and 0.652 on the test data.

3.2   Feature Importance

To get a better understanding of which brain regions drive the predictions, we inspected feature importance for each individual model. Feature importance for the linear model is readily available from the model's coefficients. The estimated coefficients of all features included in the final linear model are summarized in Table 3. Since we unified the scale of features, we can directly compare the absolute value of coefficients to obtain a ranking. Our model assigned the highest importance to the volume of the brain mask, where an increase in volume is associated with a decrease in fluid intelligence score. The volume of the right ventral diencephalon is ranked second and positively correlated with fluid intelligence.

For gradient boosting models, we computed SHAP values for each subject and feature in the training data. By ranking features according to the mean absolute SHAP value, we get an overall ranking of features according to their importance. We list the top 20 SRI features by mean absolute SHAP value ϕ in Fig. 2a. The top ranked features are the volumes of left parahippocampal gyrus ($\phi = 0.0204$) and pons white matter ($\phi = 0.0202$), while right parahippocampal gyrus ($\phi = 0.0155$) follows with a slightly lower SHAP value. We note that the maximum mean absolute SHAP value of 0.0204 for left parahippocampal gyrus is relatively small, compared the overall scale of the standardized residualized fluid intelligence score (min $= -4.1905$, max $= 3.0276$). Therefore, we do not observe a strong association between a single brain region and fluid intelligence.

Table 3. Estimated coefficients for all features used in the linear ridge regression model.

Feature	Coefficient
Volume of the brain mask	−0.284
Right Hippocampus volume	0.131
Right Ventral Diencephalon volume	0.206
Left Paracentral thickness	0.082
Right Cortical White Matter volume	0.113
Right Fusiform thickness	−0.050
Right Superior Temporal thickness	−0.009
Left Lateral Orbitofrontal thickness	−0.066

(a)

(b)

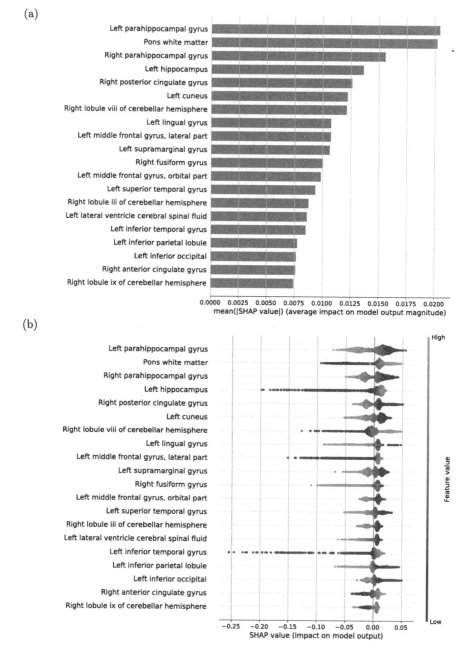

Fig. 2. Gradient boosting with SRI: (a) Top 20 features sorted by mean absolute SHAP value $\bar{\phi}_j$. (b) SHAP values of top 20 features for each subject in the training data. In each row SHAP values ϕ_j for each subject are plotted horizontally, stacking vertically to avoid overlap. Each dot is colored by the value of that feature, from low (blue) to high (red). If the impact of the feature on the model's prediction varies smoothly as its value changes then this coloring will also appear smooth. (Color figure online)

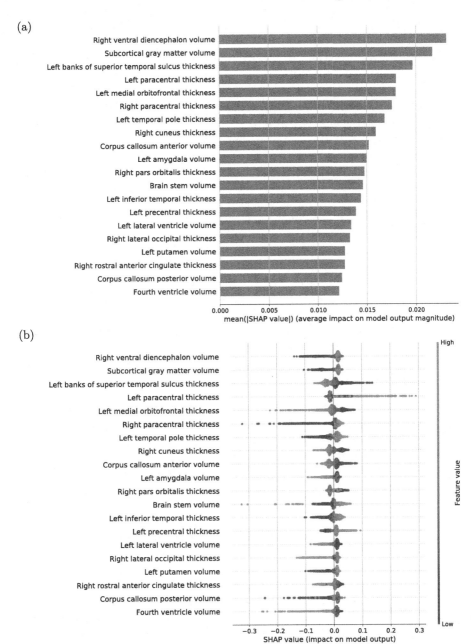

Fig. 3. Gradient boosting with FreeSurfer: (a) Top 20 features sorted by mean absolute SHAP value $\bar{\phi}_j$. (b) SHAP values of top 20 features for each subject in the training data. In each row SHAP values ϕ_j for each subject are plotted horizontally, stacking vertically to avoid overlap. Each dot is colored by the value of that feature, from low (blue) to high (red). If the impact of the feature on the model's prediction varies smoothly as its value changes then this coloring will also appear smooth. (Color figure online)

Instead, it seems that the interplay between multiple region seems to be more important.

The top 20 FreeSurfer features by mean absolute SHAP value ϕ are listed in Fig. 3a. Nine out of the top 20 features represent volume measurements, while the remainder represents thickness measurements. The top ranked feature are the volume of the right ventral diencephalon ($\phi = 0.0231$), followed by subcortical gray matter volume ($\phi = 0.0217$) and left banks of superior temporal sulcus thickness ($\phi = 0.0197$). As we observed above for SRI features, the top ranked feature still has a relatively small effect on the prediction of the model on its own; the right ventral diencephalon volume with $\phi = 0.0231$ on average only contributes a small fraction to the model's predictions. Therefore, single FreeSurfer features, too, do not have a strong association with fluid intelligence.

When comparing all three models, we note that there is little overlap in the most discriminative features. The linear model and gradient boosting with SRI features, both identified hippocampus, fusiform, and superior temporal gyrus to be associated with fluid intelligence. Rankings based on FreeSurfer features used in the linear model and gradient boosting share ventral diencephalon volume and paracentral thickness. Interestingly, while the linear model assigned the highest importance to the overall brain volume, it is absent from the top 20 features when using gradient boosting. Considering that the prediction error for all models is relatively high and that the contribution of individual features are low (small mean SHAP value), it is not surprising that we were unable to identify a clear signature of feature importance across all models considered here.

In addition to the mean absolute SHAP value, we can also look at individual, subject-specific SHAP values depicted in Fig. 2b for SRI features and Fig. 3b for FreeSurfer features. Figure 2b shows that left and right parahippocampal gyrus volume are negatively associated with fluid intelligence, meaning higher volumes result in a smaller predicted residualized fluid intelligence score. In contrast, the association for pons white matter is positive. Figure 3b for FreeSurfer features shows that higher volume of right ventral diencephalon is associated with an increase in fluid intelligence score, which is consistent with the linear model in Table 3. In addition, higher thickness of the left banks of superior temporal sulcus are associated with a decrease in fluid intelligence score.

Moreover, Fig. 2b indicates that the absolute SHAP value for volume of left inferior temporal gyrus and left hippocampus for a small number of subjects is orders of magnitude larger than that of the average patient. For these subjects, the predicted fluid intelligence score can drop by up to ≈ 0.25 when including one of these features. Thus, these volumes have a very high impact for these patients, although the global impact is modest. Similar effects can be observed in Fig. 3b for FreeSurfer features. For instance, we can observe SHAP values up to ≈ -0.3 for brain stem volume, which indicates a significantly bigger importance than the average SHAP value of 0.0146. For these subjects, the brain stem volume highly impacts the model's prediction.

Finally, non-linear effects of estimated models can be observed. Figure 2b shows that for left middle frontal gyrus (orbital part) and right lobule ix of cerebellar hemisphere, large volumes can be associated with either positive or negative SHAP values. Figure 3b shows that brain stem volume seems to be nonlinearly correlated with fluid intelligence: large volumes can be associated with either positive or negative SHAP values.

4   Conclusion

We studied the prediction of fluid intelligence from T1-weighted magnetic resonance images based on features derived from the SRI24 atlas and FreeSurfer. We proposed a linear ridge regression model with 4 volume, and 4 thickness measurements of the brain, and compared it to gradient boosting models using 122 (SRI24) and 136 (FreeSurfer) measurements of the brain, respectively. We experienced predicting fluid intelligence from MRI scans to be a generally difficult task. Our experiments showed that using FreeSurfer features offers a slight advantage in prediction accuracy over SRI24 features, and further gradient boosting over our linear approach. Our gradient boosting model with FreeSurfer ranked third place among 24 submissions to the ABCD Neurocognitive Prediction Challenge. The model indicates that both higher volumes of right ventral diencephalon and higher volumes of subcortical gray matter are associated with an increase in fluid intelligence score. At the same time, we did not find sufficient evidence that fluid intelligence is influenced by only one or a few brain regions. Therefore, we conclude that future research should focus on the interaction between different regions in the brain and their development over time to get a better understanding of which neurobiological factors could drive fluid intelligence.

Acknowledgements. This research was partially supported by the Bavarian State Ministry of Education, Science and the Arts in the framework of the Centre Digitisation.Bavaria (ZD.B).

References

1. Blair, C.: How similar are fluid cognition and general intelligence? A developmental neuroscience perspective on fluid cognition as an aspect of human cognitive ability. Behav. Brain Sci. **29**, 109–125; Discussion 125–160 (2006)
2. Bron, E.E., et al.: Standardized evaluation of algorithms for computer-aided diagnosis of dementia based on structural MRI: the caddementia challenge. NeuroImage **111**, 562–579 (2015)
3. Carroll, J.B.: Human Cognitive Abilities. Cambridge University Press, Cambridge (1993)
4. Chen, T., Guestrin, C.: XGBoost: a scalable tree boosting system. In: Proceedings of the 22nd ACM SIGKDD International Conference on Knowledge Discovery and Data Mining, pp. 785–794 (2016)
5. Ferrer, E.: Fluid reasoning and the developing brain. Front. Neurosci. **3**(1) (2009)
6. Fischl, B.: FreeSurfer. NeuroImage **62**(2), 774–781 (2012)
7. Friedman, J.H.: Greedy function approximation: a gradient boosting machine. Ann. Stat. **29**(5), 1189–1232 (2001)

8. Friedman, J.H.: Stochastic gradient boosting. Comput. Stat. Data Anal. **38**(4), 367–378 (2002)
9. Gray, J.R., Chabris, C.F., Braver, T.S.: Neural mechanisms of general fluid intelligence. Nat. Neurosci. **6**(3), 316–322 (2003)
10. Haier, R.J., Jung, R.E., Yeo, R.A., Head, K., Alkire, M.T.: Structural brain variation and general intelligence. NeuroImage **23**(1), 425–433 (2004)
11. Hoerl, A.E., Kennard, R.W.: Ridge regression: biased estimation for nonorthogonal problems. Technometrics **12**(1), 55–67 (1970)
12. Kozachenko, L.F., Leonenko, N.N.: Sample estimate of the entropy of a random vector. Problemy Peredachi Informatsii **23**(2), 9–16 (1987)
13. Lundberg, S.M., Lee, S.I.: A unified approach to interpreting model predictions. In: Advances in Neural Information Processing Systems 30, pp. 4765–4774 (2017)
14. Narr, K.L., et al.: Relationships between IQ and regional cortical gray matter thickness in healthy adults. Cereb. Cortex **17**(9), 2163–2171 (2006)
15. Pedregosa, F., et al.: Scikit-learn: machine learning in python. J. Mach. Learn. Res. **12**, 2825–2830 (2011)
16. Pfefferbaum, A., et al.: Altered brain developmental trajectories in adolescents after initiating drinking. Am. J. Psychiatry **175**(4), 370–380 (2018)
17. Pölsterl, S., Gutiérrez-Becker, B., Sarasua, I., Guha Roy, A., Wachinger, C.: An auto-ML approach for the prediction of fluid intelligence from MRI-derived features. In: Pohl, K.M., et al. (eds.) Adolescent Brain Cognitive Development Neurocognitive Prediction Challenge (ABCD-NP-Challenge), ABCD-NP 2019. LNCS, vol. 11791, pp. 99–107 (2019)
18. Rohlfing, T., Zahr, N.M., Sullivan, E.V., Pfefferbaum, A.: The SRI24 multichannel atlas of normal adult human brain structure. Hum. Brain Mapp. **31**(5), 798–819 (2010)
19. Shahriari, B., Swersky, K., Wang, Z., Adams, R.P., de Freitas, N.: Taking the human out of the loop: a review of Bayesian optimization. Proc. IEEE **104**(1), 148–175 (2016)
20. Shapley, L.S.: A value for n-person games. Contrib. Theory Games **2**(28), 307–317 (1953)
21. Shaw, P., et al.: Intellectual ability and cortical development in children and adolescents. Nature **440**(7084), 676–679 (2006)
22. Snoek, J., Larochelle, H., Adams, R.P.: Practical Bayesian optimization of machine learning algorithms. Adv. Neural Inf. Process. Syst. **25**, 2951–2959 (2012)
23. Štrumbelj, E., Kononenko, I.: Explaining prediction models and individual predictions with feature contributions. Knowl. Inf. Syst. **41**(3), 647–665 (2014)
24. Wachinger, C., Reuter, M., Alzheimer's Disease Neuroimaging Initiative, et al.: Domain adaptation for Alzheimer's disease diagnostics. Neuroimage **139**, 470–479 (2016)
25. Wright, S., Matlen, B., Baym, C., Ferrer, E., Bunge, S.: Neural correlates of fluid reasoning in children and adults. Front. Hum. Neurosci. **2**, 8 (2008)
26. Zhang, C., Liu, C., Zhang, X., Almpanidis, G.: An up-to-date comparison of state-of-the-art classification algorithms. Expert Syst. Appl. **82**, 128–150 (2017)

Ensemble of SVM, Random-Forest and the BSWiMS Method to Predict and Describe Structural Associations with Fluid Intelligence Scores from T1-Weighed MRI

Jose Tamez-Pena[✉] [iD], Jorge Orozco [iD], Patricia Sosa,
Alejandro Valdes, and Fahimeh Nezhadmoghadam

Tecnologico de Monterrey, 64849 Monterrey, NL, Mexico
jose.tamezpena@tec.mx

Abstract. The degree of association between fluid intelligence and neuroanatomy is important in refining our understanding of brain development. The primary goal of this work is twofold: to predict fluid intelligence from T1-weighed MRI, and to describe the MRI features that are associated with fluid intelligence. In this paper, we propose to ensemble the predictions of three machine learning strategies: Support Vector Machine (SVM), Random Forest (RF), and Bootstrapped Step Wise Model Selection (BSWiMS). Gender-stratified SVM was trained on children using age (ages 9–10), plus 122 volumetric scores provided by the ABCD challenge team. RF and BSWiMS were gender-stratified and trained using cubic root transformed data, summarized by left-right mean and relative absolute differences, and augmented by 19 volumetric statistical descriptors of major anatomical regions. Then, the transformed-augmented feature set was adjusted by age and the mean volume of the training set. The predictions of the three models were averaged to get the final prediction on each one of the test subjects. The Mean Squared Error (MSE) of MRI-predicted fluid intelligence on the test subjects was 100.89. The top features associated with fluid intelligence were the volume of the pons white mater and the volume of the parahippocampal gyrus.

Keywords: Fluid intelligence · Machine learning · MRI

1 Introduction

Fluid intelligence is a major factor in measuring general intelligence [1]. Therefore, the ABCD, the largest long-term study of brain development and child health in the United States, has captured MRI data as well as genetics, neuropsychological, behavioral, and other health assessments to "determine how childhood experiences (such as sports, videogames, social media, unhealthy sleep patterns, and smoking) interact with each other and with a child's changing biology to affect brain development and social, behavioral, academic, health, and other outcomes". Hence, one specific challenge is

© Springer Nature Switzerland AG 2019
K. M. Pohl et al. (Eds.): ABCD-NP 2019, LNCS 11791, pp. 47–56, 2019.
https://doi.org/10.1007/978-3-030-31901-4_6

determining if machine learning prediction methods can be used to predict fluid intelligence from T1-weighted MRI [2].

Regarding prediction methods, there are hundreds of methods that can be used to predict a continuous variable from a set of features. Random Forest [3], Support Vector Machines [4], Recursive partitioning for classification [5], and Least Absolute Shrinkage and Selection Operator (LASSO) [6] are among the main machine learning approaches that can learn a linear structure from a set of predictors. Furthermore, there are several ways to improve prediction by first selecting the features that are associated with the data. Once selected, they're used in a regression equation. Bootstrap Stage-Wise Model Selection (BSWiMS) used this strategy, plus bootstrap samples, to extract a set of regression models that predict the desired outcome [7]. The aim of this work is to explore the behavior of these approaches on the ABCD challenge data, report the features used to predict the fluid intelligence, build a final prediction based on top three predictors, and finally predict validation and testing data sets.

2   Materials and Methods

Data was provided by the ABCD challenge team: 3739, 415, and 4515 children were used for training, validation, and testing, respectively. All the children were measured by standardized protocols, and MRI features were extracted from T1-weighted images using a standard procedure described by the ABCD challenge organizers [2]. In this paper, we describe the methods used to predict the fluid intelligence score from the provided data sets.

2.1   Data Processing

We transformed all the volume scores provided via the challenge organizers by computing the cubic root on all measurements. This operation transformed the inherent distribution of volumetric errors into a Gaussian distribution and consequently mitigating the influence of measurement errors on each one of the volumes' distributions. Furthermore, brain asymmetry has been associated with several chronic neurological diseases such as Schizophrenia and Alzheimer [8, 9]; therefore, here we explore the hypothesis that structural asymmetry may also be associated with fluid intelligence. To explore this hypothesis, we computed the mean and absolute difference of the left and right volumes of each the brain structures. Finally, we augmented the set of transformed features by adding the descriptive features described in Table 1. All transformations and descriptive computations were done in Microsoft Excel (2013).

Once we computed the descriptive features, we adjusted all the features (transformed and descriptive) to any residual associations to volume or age, and we did this independently on males and females. The adjustment was done by computing the residuals of the following linear model:

$$f_i = \beta_o + \beta_1 Age + \beta_2 AVGVolume + \beta_3 Age * AVGVolume, \tag{1}$$

on all features using the FRESA.CAD package [7].

Table 1. Descriptive features derived from volumetric data. Transformed volumes were computed using the cubic root transformation.

Descriptive feature	Description
AVGVolume	Mean of all transformed volume
STDVols	Standard deviation of transformed volumes
volCV	Coefficient of Variation
AvgAbsDif	Mean value of Relative Absolute Differences
StdevDiff	Standard Deviation of Relative Absolute Differences
CVDif	The coefficient of Relative Absolute Differences
GMV	Mean value of gray matter
STDGM	The standard deviation of gray matter
GMVCV	The coefficient of variation of gray matter
Cerebellum	Mean Cerebellum Volume
STDCer	The standard deviation of Cerebellum
CerCV	The coefficient of Variation of Cerebellum
WMV	Mean value of White Matter
STDWM	The standard deviation of white matter
CVWM	The coefficient of variation of Gray Matter
GWMWMVR	Gray Matter to White matter ratio
FluidV	Mean Fluid Volume
STDFLID	The standard deviation of Fluid
FLIDCV	The coefficient of Variation of Fluid

2.2 Benchmarking Machine Learning Methods

The adjusted training data was explored by a Benchmarking function provided by the FRESA.CAD R package. The FRESA.CAD benchmarking function does a repeated hold-out cross-validation (RHOCV) [10] of the following machine learning methods: Recursive partitioning (RPART), Random Forest (RF), Support Vector Machine (SVM), LASSO, and Bootstrapped Stage-Wise Model Selection (BSWiMS) [3–6, 11]. Furthermore, the benchmarking evaluates the efficiency of different feature selection methods like minimum redundancy maximum relevance mRMR [11] and univariate selection adjusted for false discovery rate [12]. Features selected by these methods were used to train robust regression, ridge regression, and linear regression models. Trained models were evaluated in the holdout testing data. Finally, internal-test performance results were reported. Besides exploring the regression models using the augmented data, we also explored the capability of the original ABCD challenge volume scores to model fluid intelligence.

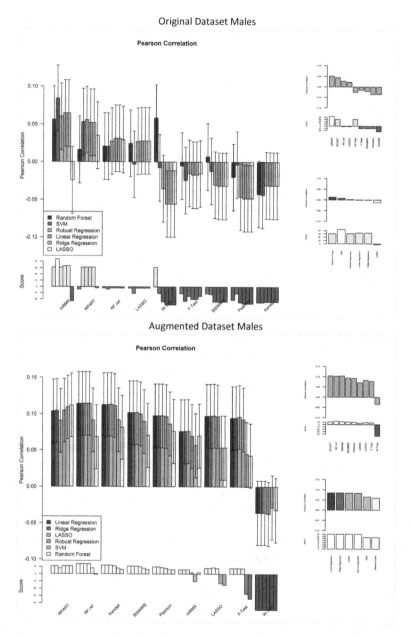

Fig. 1. Benchmarking Machine Learning Regression Algorithms for male subjects. Pearson correlation results of the predicted fluid intelligence to fluid intelligence. The top plots show the Pearson correlation of several; ML methods using the training dataset provided by the ABCD challenge. The bottom plots show the Pearson correlation of the same methods using the augmented data set.

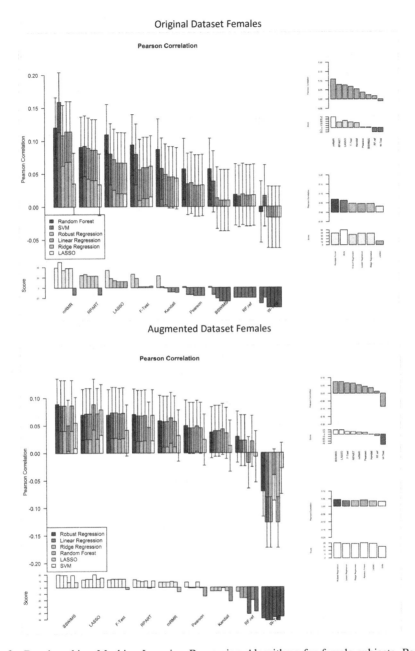

Fig. 2. Benchmarking Machine Learning Regression Algorithms for female subjects. Pearson correlation results of the predicted fluid intelligence to fluid intelligence. The top plots show the Pearson correlation of several; ML methods using the training dataset provided by the ABCD challenge. The bottom plots show the Pearson correlation of the same methods using the augmented data set.

All the benchmarks were run separately on males' and females' data set using 90% of the data for training and holding 10% of the data for test evaluation. The hold-out procedure was repeated 100 times. Figures 1 and 2 show the results of benchmarking the regression methods on the ABCD challenge data set. The Pearson correlation between fluid intelligence and benchmarking test results indicated that BSWiMS had an equivalent performance on both males' and females' data sets on the augmented training data. On the other hand, the CV results indicate that SVM was able to predict males' and females' datasets based only on features provided by the original training set.

2.3   Method Ensemble

The results of the Benchmarking function on the training data set were used to determine successful machine learning strategies that may be able to predict the fluid intelligence on the validation and testing data. We selected three methods according to the benchmark results. The first method was SVM. SVM showed a statistically significant association using original features. The second method was the BSWiMS that selected linear models composed on a small number of features ($n < 8$), and it had an equivalent performance on both males and females. Finally, we selected the Random Forest method because RF uses an independent learning methodology than SVM and BSWiMS, and the benchmarking results indicated that it was able to predict fluid intelligence. Regarding the selected modeling methods, SVM used the radial kernel, RF is based on decision trees, and BSWiMS built simple linear models. Hence, they approached the regression problem using three different strategies. Once we selected the methods, we refit the SVM, BSWiMS, and RF using the training data sets. SVM was fitted to male and female by training original data using the radial basis function, and we did not attempt to optimize any of the SVM parameters. Once trained, we predicted the validation and testing sets. For BSWiMS and RF we refit the models using the augmented training data set. First, we fitted the BSWiMS model using default parameters and all the sex-stratified training data. The BSWiMS procedure was repeated 10 times, and the final model coefficients were bagged. We also fitted the data utilizing the RF method, by using default parameters. Hence, no parameter optimization was attempted. To minimize prediction noise, we repeated the sex-stratified RF fitting 15 times. Once BSWiMS and RF were fitted on males and females, we proceeded to predict the validation and to test datasets. First by cubic-root transformation, then we augmented the features using the described features. Finally, we adjusted the data using the coefficients of Eq. 1. The final prediction was the average of the three methods.

2.4   Prediction Evaluation

The evaluation of the results was done using the ABCD challenge validation data set using the R code provided by challenge organizers. The code evaluated the mean square error (MSE) and the R^2 between predicted scores and the actual scores.

3 Results

3.1 Benchmarking Results

Figures 1 and 2 shows the Pearson correlation results of the repeated hold-out validation strategy using only validation data. The Pearson correlation varied from not being statistically significant, for BSWiMS method on the original data set, to a maximum correlation of 0.16 with 95% confidence interval of 0.12 to 0.21. Figure 3 shows the most commonly selected features by the different filtering strategies using the transformed data sets. It is clear that the features that are useful to predict fluid intelligence are different between males and females. Only five features: |frontal-sup_gm|, |wm400_wm|, insula_gm, parahippocampal_gm, and parietalinf_gm were common between males and females. Among these five features, two were related to differences between left and right volumes, and three were related to volume size.

3.2 Feature Relevance

The BSWiMS procedure returns a summary analysis of the relevant features required to model the fluid intelligence score using the sex-stratified transformed data sets. Table 2 shows the required features needed to model the male intelligence scores. Table 3 shows the features involved in predicting female scores. The tables show that the volume of the parahippocampal gyrus and the volume of the pons are an important feature associated with fluid intelligence scores in both males and females. Moreover, the other features are unique to males and females. Among them, four features are associated with differences between the left and right volumes in males while only two features associated with differences between and right volumes affect females significantly.

3.3 Models Performance

The training results of the BSWiMS procedure indicated that the linear models associated with fluid intelligence can explain 3.6% and 2.2% of the male and the female variance, respectively. Table 4 shows the MSE of the internal cross-validation of the models based. The validation results of the ensemble procedure showed that structural MRI predicted 0.7% (MSE = 71.86) of the fluid intelligence variance and that this prediction was marginally significant (p-value = 0.06). Finally, the test results indicated that the ensemble method had an MSE of 100.89.

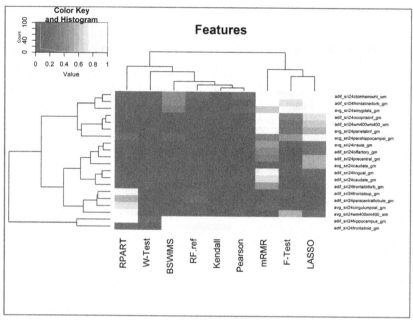

Male Feature

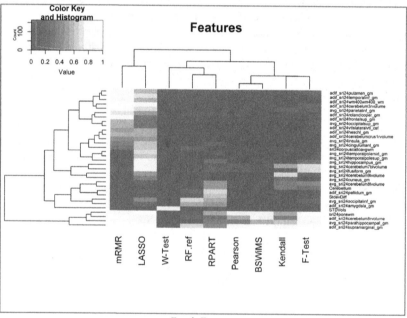

Female Features

Fig. 3. Features associated with Fluid Intelligence Scores dived by Filter Method. The heat-map shows the relative frequency of the feature selection, where yellow represents the most common discovery after 100 repetitions of the hold-out cross-validation. (Color figure online)

Table 2. Male Model. |°| are features related to the left-right absolute difference

Feature	Beta estimate	Univariate MSE	F test p-value
|hippocampus_gm|	−53.37 (−76.93 to −29.82)	90.9	4.48E−06
|frontalmid_gm|	−34.21 (−50.71 to −17.70)	91.1	2.44E−05
amygdala_gm	2.58 (1.04 to 4.13)	91.5	5.12E−04
parahippocampal_gm	−1.45 (−2.37 to −0.54)	91.5	9.59E−04
ponswm	0.91 (0.30 to 1.51)	91.3	1.60E−03
|frontalmedorb_gm|	−13.19 (−23.15 to −3.23)	91.5	4.73E−03
|cblmhemiwht_wm|	81.38 (19.24 to 143.50)	91.5	5.13E−03

MSE: *88.43*
R2: *0.03671*

Table 3. Female Model. |°| are features related to the left-right absolute difference

Feature	Beta estimate	Univariate MSE	F test p-value
ponswm	1.09 (0.49 to 1.69)	79.0	1.90E−04
|supramarginal_gm|	−13.18 (−21.18 to −5.19)	79.0	6.18E−04
|cerebelum8rvolume|	−25.86 (−42.81 to −8.91)	79.0	1.39E−03
parahippocampal_gm	−1.23 (−2.13 to −0.33)	79.1	3.60E−03

MSE: *77.72*
R2: *0.02226*

Table 4. MSE of the internal validation of the three models

Gender	BSWiMS	RF	SVM
Males	91.0	91.2	95.0
Females	79.0	78.6	83.3

4 Discussion

Machine learning strategies can be used to explore systematically different methodologies to model and predict abstracts outcomes like fluid intelligence. The exploration of volumetric data and transformed data showed that data transformation can be useful in finding anatomical structures that may affect the expression or development of child fluid intelligence. Several brain structures were found to explain the difference in the intelligence score, and the aggregation of them in linear models increased the power to describe the difference between children. We also found that differences between the left and right structures in the brain are associated with the fluid intelligence scores. Clear gender differences were also observed. Males were modeled by simple linear models while explaining the female fluid intelligence required pattern matching methods. In other words, there are patterns in female fluid intelligence that are difficult to model by simple linear models. Even with this difficulty, we found that there are significant associations between changes in the pons white mater, parahippocampal gyrus, and fluid intelligence in both males in females and that this association can only be explained 0.7% of the variance. Hence, the reported findings are indicating that there are factors beyond differences in brain structures that play a more important role in the development of child fluid intelligence.

5　Conclusion

Machine learning methods can be used to discover the main factors associated with fluid intelligence. Here, we found that brain morphology plays a small role in explaining the difference in fluid intelligence in children between ages 9 and 10.

References

1. Carroll, J.B.: Human Cognitive Abilities by John B. Carroll (2019)
2. Pohl, K.M., Thompson, W.K.: https://sibis.sri.com/abcd-np-challenge/
3. Breiman, L.: Random forests. Mach. Learn. **45**, 5–32 (2001)
4. Cortes, C., Vapnik, V.: Support-vector networks. Mach. Learn. **20**, 273–297 (1995)
5. De'ath, G., Fabricius, K.E.: Classification and regression trees: a powerful yet simple technique for ecological data analysis. Ecology **81**, 3178–3192 (2000)
6. Tibshirani, R.: Regression shrinkage and selection via the Lasso. J. R. Stat. Soc. Ser. B-Methodol. **58**, 267–288 (1996)
7. Tamez-Pena, J.G., Tamez-Pena, M.J.G.: Package 'FRESA. CAD' (2014)
8. Toga, A.W., Thompson, P.M.: Mapping brain asymmetry. Nat. Rev. Neurosci. **4**, 37–48 (2003)
9. Thompson, P.M., et al.: Cortical variability and asymmetry in normal aging and Alzheimer's disease. Cereb. Cortex **8**, 492–509 (1998)
10. Kim, J.H.: Estimating classification error rate: repeated cross-validation, repeated hold-out and bootstrap. Comput. Stat. Data Anal. **53**, 3735–3745 (2009)
11. Ding, C., Peng, H.: Minimum redundancy feature selection from microarray gene expression data. J. Bioinform. Comput. Biol. **3**, 185–205 (2005)
12. Benjamini, Y., Hochberg, Y.: Controlling the false discovery rate - a practical and powerful approach to multiple testing. J. R. Stat. Soc. Ser. B-Methodol. **57**, 289–300 (1995)

Predicting Intelligence Based on Cortical WM/GM Contrast, Cortical Thickness and Volumetry

Juan Miguel Valverde[1](✉) ⓘ, Vandad Imani[1](✉) ⓘ, John D. Lewis[2](✉) ⓘ,
and Jussi Tohka[1](✉) ⓘ

[1] AI Virtanen Institute for Molecular Sciences, University of Eastern Finland,
Kuopio, Finland
{juanmiguel.valverde,vandad.imani,jussi.tohka}@uef.fi
[2] Montreal Neurological Institute, McGill University, Montreal, Canada
jlewis@bic.mni.mcgill.ca

Abstract. We propose a four-layer fully-connected neural network
(FNN) for predicting fluid intelligence scores from T1-weighted MR
images for the ABCD-challenge. In addition to the volumes of brain
structures, the FNN uses cortical WM/GM contrast and cortical thickness at 78 cortical regions. These last two measurements were derived
from the T1-weighted MR images using cortical surfaces produced by
the CIVET pipeline. The age and gender of the subjects and the scanner manufacturer are also used as features for the learning algorithm.
This yielded 283 features provided to the FNN with two hidden layers of
20 and 15 nodes. The method was applied to the data from the ABCD
study. Trained with a training set of 3736 subjects, the proposed method
achieved a MSE of 71.596 and a correlation of 0.151 in the validation set
of 415 subjects. For the final submission, the model was trained with
3568 subjects and it achieved a MSE of 94.0270 in the test set comprised
of 4383 subjects.

Keywords: Artificial neural networks · Machine learning · Magnetic
resonance imaging · Fluid intelligence · Cortical thickness ·
Cortical contrast

1   Introduction

Fluid intelligence is a core factor of general intelligence. The rate at which skills
and knowledge, i.e. crystallized intelligence, are acquired depends upon it. Thus,
there is great interest in determining the extent to which fluid intelligence can
be determined from brain measures.

In this study, we used a supervised learning model to automatically predict fluid intelligence scores, *i.e.* fluid intelligence with demographic confounding factors removed, based on T1-weighted Magnetic Resonance Images (MRIs)

J. M. Valverde and V. Imani—Contributed equally to this work.

© Springer Nature Switzerland AG 2019
K. M. Pohl et al. (Eds.): ABCD-NP 2019, LNCS 11791, pp. 57–65, 2019.
https://doi.org/10.1007/978-3-030-31901-4_7

at 3T. More specifically, this report describes our submission to the the ABCD Neurocognitive Prediction Challenge (ABCD-NP-Challenge 2019)[1]. Data were obtained from the NIMH Data Archive (NDA) database[2], generated by the Adolescent Brain Cognitive Development (ABCD) study, the largest long-term study of brain development and child health in the United States. The feature set used for the prediction model included sociodemographic (age, gender) and MRI-derived measures. Our MRI-derived features included regionally averaged cortical thickness and white/grey contrast measures in addition to the volumes of a set of regions of interest provided by the challenge organizers. Relying on this feature set, we create a supervised regression framework utilizing a four-layer fully-connected neural network (FNN) to predict fluid intelligence scores.

2   Material

2.1   Training and Test Data

T1-weighted MR images, volumetric measures, age, gender, scanner and fluid intelligence scores were available to the ABCD challenge participants via National Database for Autism Research (NDAR) website. A primary consideration in measuring general intelligence is the role of fluid intelligence [5] which was measured via the NIH Toolbox Neurocognition battery [2] and from which demographic factors (e.g., sex and age) are eliminated to remove the effect of confounding variables. For this residualization, the challenge organizers used all subjects without any missing values in the data collection site, sociodemographic variables and brain volume to build a linear regression model. This model was constructed with fluid intelligence as the independent variable and the other attributes as dependent variables. The residuals computed for all subjects provided the fluid intelligence scores to be predicted.

From a total of 8553 individual subjects, fluid intelligence scores were provided for participants in the training set (3736 subjects) and validation set (415 subjects), whereas the other subjects (4402 subjects) formed the test set. As explained in Sect. 2.2, few of these subjects could not be processed through CIVET pipeline, hence they were not used to predict the fluid intelligence scores. The age and gender characteristics of the 8347 subjects across scanners from three different manufacturers used in this work are presented in Table 1. There was no significant difference in age and gender between different scanners.

The FNN was trained with a set of 3568 subjects and validated during the training process on the validation set consisting of 396 subjects. Afterwards, the proposed method was used to generate the predictions of the fluid intelligence scores of the 4383 subjects in the test set.

---

[1] https://sibis.sri.com/abcd-np-challenge/.

[2] https://nda.nih.gov.

Table 1. The subject characteristic of this work. The ages of subjects ranged from 107 to 133 months.

Scanners	GE (25%)			SIEMENS (60%)			Philips (15%)		
	Train	Validation	Test	Train	Validation	Test	Train	Validation	Test
Subjects	966	99	1029	2101	240	2698	501	57	656
Females	48.2%	47.4%	49.1%	46%	51.6%	47.3%	49%	45.6%	49.2%

2.2   Image Pre-processing

Volume measures, provided by the competition organizers, were derived from the T1-weighted images as follows: the Minimal Processing pipeline [8] transformed the raw data into NIfTI format. Afterwards, the NCANDA pipeline [16] defined a brain mask by non-linearly mapping the SRI24 atlas [17] to the T1-weighted images, and it removed noise and corrected for bias-field inhomogeneities. Several skull-stripping methods were used with bias-field corrected and non-bias-field corrected images, and a majority voting of the resulting masks refined the brain masks from the previous step. The skull-stripped brains were newly corrected for bias-field inhomogeneities and segmented into gray matter, white matter and cerebrospinal fluid via Atropos [4]. The SRI24 atlas was non-rigidly registered to the images via ANTS [3] to further parcellate the gray matter and the resulting segmentations were linearly registered to the SRI24 atlas. Finally, results that failed to pass a visual two-tier quality check were rejected.

In addition to volume measures, we used cortical thickness and cortical white/gray contrast measures that were regionally averaged based on the Automated Anatomical Labeling (AAL) atlas [21]. For this, the T1-weighted volumes were denoised [15] and processed with CIVET (version 2.1 ; 2016), a fully automated structural image analysis pipeline developed at the Montreal Neurological Institute[3]. CIVET corrects intensity non-uniformities using N3 [18]; aligns the input volumes to the Talairach-like ICBM-152-nl template [7]; classifies the image into white matter, gray matter, cerebrospinal fluid, and background [20,22]; extracts the white-matter and pial surfaces [11]; and maps these to a common surface template [14].

Cortical thickness was measured in native space at 81924 vertices using the Laplacian distance between the two surfaces. The Laplacian distance is the length of the path between the gray and white surfaces following the tangent vectors of the cortex represented as a Laplacian field [10]. The CT measures were averaged into 78 regional measures relying on the AAL atlas.

To extract the white/gray contrast measures, similarly to [13], the intensity on the T1-weighted MRI was sampled 1 mm inside and 1 mm outside of the white surface, and the ratio of the two measures was formed. Here we used a highly simplified version of the algorithm of [13] and generated supra-white and sub-white surfaces relying on the surface normals provided by the CIVET pipeline. The intensity values on the T1-weighted image (without non-uniformity correction

---

[3] http://www.bic.mni.mcgill.ca/ServicesSoftware/CIVET-2-1-0-Introduction.

or normalization) were sampled at each vertex of both the supra-white surface and the sub-white surface, and the ratio was formed by dividing the value at each vertex of the sub-white surface by the value at the corresponding vertex of the supra-white surface. Similarly to CT measures, the contrast measures were averaged into 78 regional measures relying on the AAL atlas. The white/gray contrast measures are sensitive to scanner-specific differences in tissue contrast [13], so to correct for this, we normalized the contrast values per scanner manufacturer by z-scoring the contrast values scanner manufacturer-wise as explained in detail in [13].

The CIVET pipeline failed to process 168 subjects from the training set, 19 subjects from the validation set and 19 subjects from the test set most likely due to motion artifacts and/or excessive noise interfering with registration and segmentation. Consequently, a different model trained exclusively on the provided volumetric and sociodemographic data was used to infer the fluid intelligence score of the validation subjects whose derived data could not be produced.

3   Machine Learning Approach

The developed regression model based on artificial neural networks was trained with feature vectors that incorporated 122 volumetric, 78 contrast and 78 CT measures along with gender, age and the scanner manufacturer one-hot encoded. Age and image-derived attributes were normalized feature-wise by subtracting their mean and dividing them by their standard deviation. The network trained was a four-layer FNN. The model was trained using mini-batches of size 24 and Adam [12], a stochastic gradient descent method with an adaptive learning rate starting from $\eta = 0.00001$. The cost function to minimize was the mean squared error (MSE). After every epoch the model was validated and the training stopped when the MSE was greater than the minimum MSE obtained in the previous iterations plus 0.7. This stopping criteria was empirically found to increase the correlation coefficients of the predictions.

A four-layer FNN was trained by adjusting its parameters using the backpropagation algorithm to minimize the MSE produced between the desired output and the network prediction. The input layer consisted of 283 nodes corresponding to each of the features of the input vector. The two hidden layers have 20 and 15 nodes, respectively. Since predicting fluid intelligence scores is a single-output regression problem, the output layer consisted of a single node. As the network was fully connected, all nodes between successive layers were connected. This configuration provided the best results from the variations that were tried in the limited time available.

The weights of the FNN were randomly initialized as proposed in [9] whereas the bias terms were initialized to zero. An exponential linear unit (ELU) nonlinear activation function [6] was used in all intermediate layers with $\alpha = 1$ such that if $x > 0$ then $f(x) = x$, otherwise $f(x) = exp(x) - 1$. All layers except for the input layer had a dropout of rate 0.5 [19]. The training of the FNN was implemented with TensorFlow [1].

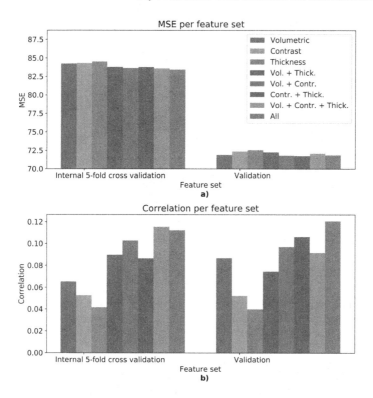

Fig. 1. Comparison of the MSE and correlation among FNNs trained on different feature sets in internal 5-fold cross-validation and validation set experiments. "All" combines volumetric, contrast and thickness measures with age, gender and the scanner manufacturer.

4   Results

Our predictions were generated on the regression model trained on the 3568 subjects for which all image-derived features were obtained. The final submission consisted of 4383 predictions achieving a MSE of 94.0270.

The performance of the model on the validation set and in an internal 5-fold cross-validation test was used to study how different combinations of features contribute to the MSE and correlation (Fig. 1). The correlation here refers to Pearson correlation coefficient between the predicted and actual fluid intelligence scores. Both experiments showed that the combination of all features provided with larger correlation coefficients (Fig. 1b). But, the variability of the MSE produced among different models trained on different set of features was small (Fig. 1a).

The proposed method trained with all features combined achieved a MSE of 81.89 and a correlation of 0.13 in an internal 5-fold cross-validation experiment. The results obtained on the validation set were a MSE of 71.596 and a correlation of 0.151, and its corresponding training loss was 84.28 (Fig. 2(a)).

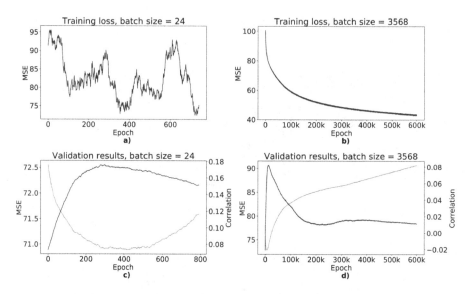

Fig. 2. Training loss and validation results of models trained with batch sizes of 24 and 3568 assessed on the validation set. Training loss is presented as moving average using a window size of 51 epochs.

As depicted in Figs. 2(a) and (b), choosing a small batch size to train the FNN causes large oscillations in the training loss whereas choosing the largest possible batch size leads to a steady decrease of the MSE. Providing with the entire training data set to train a neural network makes more accurate estimates of the gradient directions to minimize the loss over the training data. On the other hand, small batch sizes may need more iterations to converge, but the consequent fluctuations that occur during the training can lead to reaching other local minima with potentially better generalization capabilities. Figures 2(c) and (d) show that when the batch size is 3568 not only the number of required epochs was significantly larger but also the proposed FNN provided with worse MSE and correlation than when the batch size is 24.

4.1 Computation Time

The four-layer FNN was implemented in Tensorflow (Python). The total running time for training the model using all features was approximately 3 and a half minutes. Generating predictions of the testing set took half a second. The regression model was run on Ubuntu 16.04 with an Intel Xeon W-2125 CPU @ 4.00 GHz processor and 64 GB of memory.

Processing a subject through the CIVET pipeline entailed a computational time of approximately 10 hours on a cluster with Intel Xeon E5-2683 V4 @ 2.1 GHz processors using a single core with 4 GB of memory.

5   Discussion

We presented a method based on artificial neural networks to predict fluid intelligence scores from T1-weighted MR images, age and gender. Contrast and CT measures were additionally derived from the MR images to complement the provided volumetric and sociodemographic data. Training the proposed model with the combination of all image-derived features provided larger correlation coefficients than when it was trained solely on volumetric features. Nonetheless, the overall MSE of the predicted scores did not improve. The selected batch size to train the FNN caused the training loss to oscillate. However, as shown in Sect. 4, it increased the capability of the model to generalize. With this setting, the regression model achieved a MSE of 71.596 and a correlation of 0.151 in the validation set of 415 subjects. Due to the inherent complexity of the regression problem and the incorporation of additional image-derived features, future work is required to explore different architectures and deeper neural network models.

Acknowledgements. This research has been supported by grant 316258 from the Academy of Finland (to JT) and it is co-funded by Horizon 2020 Framework Programme of the European Union (Marie Skłodowska Curie grant agreement No 740264). The research also benefited from computational resources provided by Compute Canada (www.computecanada.ca) and Calcul Quebec (www.calculquebec.ca).

Data used in the preparation of this article were obtained from the Adolescent Brain Cognitive Development (ABCD) Study (https://abcdstudy.org), held in the NIMH Data Archive (NDA). This is a multisite, longitudinal study designed to recruit more than 10,000 children age 9–10 and follow them over 10 years into early adulthood. The ABCD Study is supported by the National Institutes of Health and additional federal partners under award numbers U01DA041022, U01DA041025, U01DA041028, U01DA041048, U01DA041089, U01DA041093, U01DA041106, U01DA041117, U01DA041120, U01DA041134, U01DA041148, U01DA041156, U01DA041174, U24DA041123, and U24DA041147. A full list of supporters is available at https://abcdstudy.org/nih-collaborators. A listing of participating sites and a complete listing of the study investigators can be found at https://abcdstudy.org/principal-investigators.html. ABCD consortium investigators designed and implemented the study and/or provided data but did not necessarily participate in analysis or writing of this report. This manuscript reflects the views of the authors and may not reflect the opinions or views of the NIH or ABCD consortium investigators.

The ABCD data repository grows and changes over time. The ABCD data used in this report came from doi:10.15154/1503213 (Train set); doi:10.15154/1503306 (Validation set); doi:10.15154/1503307 (Test set).

References

1. Abadi, M., et al.: TensorFlow: Large-scale machine learning on heterogeneous systems (2015). http://tensorflow.org/
2. Akshoomoff, N., Beaumont, J., Bauer, P., et al.: NIH toolbox cognition battery (CB): composite scores of crystallized, fluid, and overall cognition. Monogr. Soc. Res. Child Dev. **78**(4), 119–132 (2013). https://doi.org/10.1111/mono.12038

3. Avants, B.B., Epstein, C.L., Grossman, M., Gee, J.C.: Symmetric diffeomorphic image registration with cross-correlation: evaluating automated labeling of elderly and neurodegenerative brain. Med. Image Anal. **12**(1), 26–41 (2008)
4. Avants, B.B., Tustison, N.J., Wu, J., Cook, P.A., Gee, J.C.: An open source multivariate framework for n-tissue segmentation with evaluation on public data. Neuroinformatics **9**, 381–400 (2011)
5. Carroll, J.: Human Cognitive Abilities: A Survey of Factor-Analytic Studies, 1st edn. Cambridge University Press, Cambridge (1993). https://doi.org/10.1017/CBO9780511571312
6. Clevert, D.A., Unterthiner, T., Hochreiter, S.: Fast and accurate deep network learning by exponential linear units (ELUs). arXiv: abs/1511.07289 (2015)
7. Collins, D.L., Neelin, P., Peters, T.M., Evans, A.C.: Automatic 3D intersubject registration of MR volumetric data in standardized Talairach space. J. Comput. Assist. Tomogr. **18**(2), 192–205 (1994)
8. Hagler, D.J., et al.: Image processing and analysis methods for the adolescent brain cognitive development study. bioRxiv (2018). https://doi.org/10.1101/457739, https://www.biorxiv.org/content/early/2018/11/04/457739
9. He, K., Zhang, X., Ren, S., Sun, J.: Delving deep into rectifiers: Surpassing human-level performance on ImageNet classification. In: 2015 IEEE International Conference on Computer Vision (ICCV), pp. 1026–1034 (2015)
10. Jones, S.E., Buchbinder, B.R., Aharon, I.: Three-dimensional mapping of cortical thickness using Laplace's equation. Hum. Brain Mapp. **11**(1), 12–32 (2000)
11. Kim, J.S., et al.: Automated 3-D extraction and evaluation of the inner and outer cortical surfaces using a laplacian map and partial volume effect classification. Neuroimage **27**(1), 210–221 (2005)
12. Kingma, D.P., Ba, J.: Adam: a method for stochastic optimization. arXiv: abs/1412.6980 (2015)
13. Lewis, J.D., Evans, A.C., Tohka, J., Group, B.D.C., et al.: T1 white/gray contrast as a predictor of chronological age, and an index of cognitive performance. NeuroImage **173**, 341–350 (2018)
14. Lyttelton, O., Boucher, M., Robbins, S., Evans, A.: An unbiased iterative group registration template for cortical surface analysis. Neuroimage **34**(4), 1535–1544 (2007)
15. Manjón, J.V., Coupé, P., Martí-Bonmatí, L., Collins, D.L., Robles, M.: Adaptive non-local means denoising of MR images with spatially varying noise levels. J. Magn. Reson. Imaging **31**(1), 192–203 (2010)
16. Pfefferbaum, A., et al.: Altered brain developmental trajectories in adolescents after initiating drinking. Am. J. Psychiatry **175**(4), 370–380 (2018)
17. Rohlfing, T., Zahr, N.M., Sullivan, E.V., Pfefferbaum, A.: The SRI24 multichannel atlas of normal adult human brain structure. Hum. Brain Mapp. **31**(5), 798–819 (2010)
18. Sled, J.G., Zijdenbos, A.P., Evans, A.C.: A nonparametric method for automatic correction of intensity nonuniformity in MRI data. IEEE Trans. Med. Imaging **17**(1), 87–97 (1998)
19. Srivastava, N., Hinton, G., Krizhevsky, A., Sutskever, I., Salakhutdinov, R.: Dropout: a simple way to prevent neural networks from overfitting. J. Mach. Learn. Res. **15**, 1929–1958 (2014). http://jmlr.org/papers/v15/srivastava14a.html
20. Tohka, J., Zijdenbos, A., Evans, A.: Fast and robust parameter estimation for statistical partial volume models in brain MRI. Neuroimage **23**(1), 84–97 (2004)

21. Tzourio-Mazoyer, N., et al.: Automated anatomical labeling of activations in spm using a macroscopic anatomical parcellation of the MNI MRI single-subject brain. Neuroimage **15**(1), 273–289 (2002)
22. Zijdenbos, A.P., Forghani, R., Evans, A.C.: Automatic "pipeline" analysis of 3-D MRI data for clinical trials: application to multiple sclerosis. IEEE Trans. Med. Imaging **21**(10), 1280–1291 (2002)

Predict Fluid Intelligence of Adolescent Using Ensemble Learning

Huijing Ren, Xuelin Wang, Sheng Wang, and Zhengwu Zhang[✉]

University of Rochester, Rochester, NY 14642, USA
{Huijing_Ren,Xuelin_Wang,Sheng_Wang,Zhengwu_Zhang}@URMC.Rochester.edu

Abstract. Ensemble learning aggregates a set of models to solve the same problem and usually gives better results than a single model. We apply the ensemble method to seek a better prediction in the Adolescent Brain Cognitive Development Neurocognitive Prediction Challenge (ABCD-NP-Challenge). We manage to obtain a much better predicting accuracy on the fluid intelligence with the proposed ensemble method using volumetric data from T1w brain image than a single prediction model. In addition, we compare the results of adolescents with young adults using data from the Human Connectome Project (HCP). We find that raw fluid intelligence scores in HCP without regressing out covariates such as age and brain volume can be much better predicted by brain structure. Also, the prediction, in general, is more accurate in young adults than adolescents.

Keywords: Ensemble learning · Fluid intelligence · Brain volumetric data

1 Introduction

The Adolescent Brain Cognitive Development Neurocognitive Prediction Challenge (ABCD-NP-Challenge 2019) aims at exploring how much of fluid intelligence scores measured via the NIH Toolbox Neurocognition battery can be predicted from T1-weighted (T1w) brain imaging data. About 8,500 subjects, aging from 9 to 11 years, are provided in this challenge, separated into training data (3739 subjects), validation data (415) and testing data (4393). The fluid intelligence scores provided have been pre-processed, adjusting for brain volume, data collection site, age at baseline, sex at birth, race/ethnicity, highest parental education, parental income, and parental marital status. Volumetric data for each subject are also provided according to the segmentation results with the SRI 24 atlas [1]. In addition to these summary scores, the organizers also provide skull stripped images and raw T1w images.

H. Ren and X. Wang—Equally contribute to this paper.

Supported by the Health Sciences Center for Computational Innovation (HSCCI) at the University of Rochester.

K. M. Pohl et al. (Eds.): ABCD-NP 2019, LNCS 11791, pp. 66–73, 2019.
https://doi.org/10.1007/978-3-030-31901-4_8

The goal of this challenge is to predict fluid intelligence scores using features extracted from T1 images. Instead of re-processing the imaging data to obtain a different set of features, we rely on the provided volumetric features (122 of them) only. We propose to use ensemble learning to aggregate results from different regression models to obtain better prediction results. There are roughly two classes of ensemble methods - sequential ones and parallel ones. The sequential ensemble methods generate base learners sequentially. AdaBoost [2] is a famous example of sequential ensemble method. On the other hand, parallel ensemble methods generate independent base learners and reduce error by averaging. We take the later one to better combine some powerful base learners in this challenge.

Motivated by more comprehensively understanding the relationship between fluid intelligence and brain structure, we analyze data in another large brain imaging study, the Human Connectome Project (HCP). Different from ABCD, HCP focuses on young adults aging from 21 to 36 years. Both T1w brain image and fluid intelligence were collected in HCP, facilitating a fair comparison with the ABCD. We will apply the same model to both data sets and compare the prediction power of brain structure to fluid intelligence at different age ranges.

2 Methodology

2.1 Pre-processing

ABCD-NP-Challenge Data: In the challenge, the fluid intelligence score is processed by adjusting for many covariates, such as brain volume, data collection site, age at baseline, sex at birth, race/ethnicity, and the raw fluid intelligence score is not available. NCANDA pipeline [3] is used to pre-process the T1w brain imaging data, which includes noise removal, correction of field inhomogeneity and brain extraction. The skull-stripped T1w image is segmented into different brain tissues and the gray matter tissue is further parcellated using SRI24 atlas [1]. Volume of each brain region is calculated. The volumetric data and fluid intelligence scores have very different values due to pre-processing. We rescale the volumetric data to reduce them to the same range between 0 and 1 before model fitting.

HCP Data: We divide the HCP data (1106 subjects) into training data (885) and validation data (221). The fluid intelligence score in HCP is measured via Penn Matrix Test (PMAT), which is developed by Gur and colleagues [5]. The fluid intelligence scores are also adjusted for similar features (age, gender at birth, brain volume and so on) under the same method as ABCD-NP-Challenge Data. The T1 image is processed by Freesurfer [4]. We also employ the same rescaling method to keep the processing procedure same as the ABCD-NP-Challenge data.

2.2 Ensemble Learning

In the field of statistics and machine learning, ensemble methods have been widely used to solve different tasks. This kind of methods uses a combination

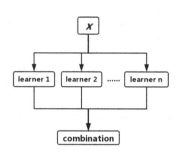

Fig. 1. The common ensemble learning structure.

Table 1. Table of a general ensemble learning algorithm.

Input:
Data set $D = \{(y_n, \mathbf{x_n}), n = 1, ..., N\}$
Algorithms of base learning $B_i, i = 1, ..., m$
Algorithm of combination method T
Process:
1.for $i = 1, ..., m,$
2.generate m base learners $B_i(\mathbf{x_n}),$
3.end.
Output:
$\hat{y_n} = T(B_1(\mathbf{x_n}), B_2(\mathbf{x_n}), ..., B_m(\mathbf{x_n}))$

of multiple learners to obtain an optimal predictive performance that is usually better than using an individual learner alone. Figure 1 shows general steps of ensemble learning, where X is the training data. The learners in the middle row are called base learners, which can be decision trees, neural networks and so on. In the last row, we call the method used to combine base learners as the top learner. For base learners, they can either be produced by the same learning method (homogeneous ensembles) or different kinds of learning algorithms (heterogeneous ensembles). Examples of homogeneous ensembles include bagging and boosting methods. To handle heterogeneous ensembles, we often adopt combination techniques such as stacking, averaging and voting. The principles of constructing an effective ensemble method are to have accurate and uncorrelated base learners [6] Table 1.

Bagging: Bagging is the abbreviation of **B**ootstrap **Agg**regat**ing**, which is a method to generate better predictor using multiple versions of a base learner [7]. The main motivation of this method is to decrease the error with the help of independent base learners. Briefly, for a data set D with N observations, sampling with replacement is used to produce m new data sets $D_i, i = 1, ..., m,$ and the sample size of each is also n. The m new data sets will be used to construct m base learners. It is clear that bagging is a parallel ensemble method. Then, voting (for classification) and averaging (for regression) can be used to combine them and obtain a more comprehensive predictor.

Boosting: Boosting is another way to create a strong learner with an ensemble of weak ones. Compared with bagging, base learners of boosting are generated differently. Here base learners are trained sequentially so that later learners can be trained based on the results of the former learners with more focus on minimizing the errors. We first train the initial base learner B_1 with the data set D_1 generated by sampling with replacement and calculate the errors $Errors(y_j, B_1(\mathbf{x_j}))$ for $(y_j, \mathbf{x_j}) \in D_1$. Then, for $\{(y_j, \mathbf{x_j})\}$ with large errors, they will be assigned a high weight to be sampled into the new data set D_2 which is used to generate the next base learner B_2. After generating a sequence of base learners, we combine these learners to get a final learner. A famous example is AdaBoost [2], which usually has good performance in classification problems. Other examples include Gradient Boosting [8] which views the boosting problem as an optimization problem and XGboost [9] (eXtreme Gradient Boosting) which is an efficient and scalable implementation of Gradient Boosting.

Combination Methods: In the final step of the ensemble, a combination method is employed to aggregate a set base learners. Dietterich [10] discussed three fundamental reasons that we can still construct a good learner using base learners with relatively high error rates. The first reason is that the combination can avoid the risk of having low prediction power of future data under one hypothesis of the training data, while the hypothesis space may be much more complicated. The second reason is that even though we have sufficient training data set, we still may not be able to find the best hypothesis computationally when performing some forms of local search for the local optima. The combination methods would provide a better approximation to the best hypothesis by running the local search from many different starting points. The third reason is related to the representation. In most applications of machine learning, the unknown hypothesis is not within the hypothesis space. The combination methods may expand the space of representable functions that may help approximate the true unknown hypothesis.

Voting, averaging and stacking are the three major combination methods [6]. With voting, the final result comes from the majority vote among the base learners, which is more popular for classification problems. With averaging, the output comes from the averaged outputs. In stacking, a new top learner is trained to combine all the outputs of the base learners. After training the base learners $B_i(\mathbf{x_j}), i = 1, ..., m$, one can use the outputs of the base learners as the input features and the original response as the input response to train the top learner, that is $y_j = T(B_1(\mathbf{x_j}), ..., B_m(\mathbf{x_j}))$. Here only the outputs of the base learners are used to reduce the risk of over-fitting.

2.3   Algorithm Used in the Challenge

Our goal is to predict the pre-processed fluid intelligence scores from T1w brain images: $y_i = f(\mathbf{x}_i)$, where y_i is the pre-processed fluid intelligence score for the subject i, $\mathbf{x}_i$ is features extracted from his/her T1w brain image and f maps from $\mathbf{x}_i$ to y_i. We will use the volumetric data as $\mathbf{x}_i$, and work on improving f. We utilize the stacking method to ensemble a set of heterogeneous base learners.

For the base learners, we first explore the simple models, such as simple linear regression model (LM) and ridge regression model. We also explore more complex models such as random forest (RF), envelope-based reduced-rank regression (RRE) [11], least absolute shrinkage and selection operator (LASSO) and Elastic-Net regression (E-Net). The other model we consider is the K-Nearest Neighbors algorithm (KNN) since it does not make assumptions on the distribution of data. These basic models will provide us a rough impression about the data set and intuition for constructing the final ensemble learning.

After training these basic models separately, correlations between predicting results of different models are calculated as a reference to pick base learners. A combination of less correlated base learners usually gives a better final model. Under this principle, we generate several groups of base learners and stack those base learners with a top learner.

In this challenge, we use mean squared error (MSE) on the validation data set as the criterion to select the best 'top + base' combination. The model with the smallest MSE would be picked out as the final ensemble learning method. To verify whether the selected model indeed can get a better prediction, we compare the stacking method with bagging, boosting and averaging using the same data.

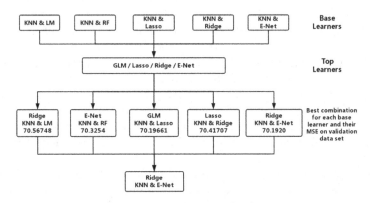

Fig. 2. The procedure of stacking.

3   Results

3.1   ABCD Results

We start from the seven basic models and the results in terms of MSE are shown in Fig. 3. We calculate the correlations between their prediction results and set *correlation* ≤ 0.2 as the criterion to select relatively uncorrelated ones as base learners of stacking and averaging methods. Figure 2 shows the stacking procedure, where the third row shows stacking combinations which perform best (with the lowest MSE) under the same base learners. According to the result, the stacking method with base learners KNN and E-Net, top learner ridge regression is most suitable to predict fluid intelligence scores in the ABCD-NP-Challenge.

Figure 3 shows the MSE of basic models and the ensemble methods applied to the ABCD-NP-Challenge data set. It is clear that most of ensemble methods indeed can improve the prediction, especially the stacking. We achieve MSE 70.19 with stacking of KNN and E-Net with top model ridge regression in the validation data set and MSE 92.99 on the testing data set.

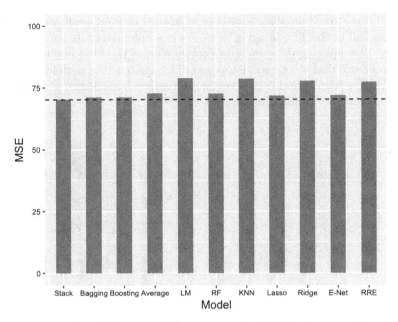

Fig. 3. The MSE from different models in terms of validation data set. The dash line is the MSE value of the stacking method.

3.2 ABCD VS HCP

To further understand the relationship between T1w brain image and fluid intelligence, we apply the same model to another big brain imaging data set, the Human Connectome Project (HCP). We compare the results between the ABCD and HCP to see whether we can replicate the finding of adolescents in adults and whether age will affect the relationship between brain structure and fluid intelligence. Same models are applied and the results are shown in Fig. 4. Since the MSE might rely on the absolute value of the original response variables, we use correlations between predicted fluid intelligence scores and measured values for evaluation. For the HCP data, we have raw fluid intelligence scores and processed scores (after regressing out some covariates). We observe that correlations from the HCP (both the raw one and the processed one) in general are better than the results from the ABCD. Especially, the raw scores can be much better predicted. We also notice that correlation values of the ensemble models are higher than those of basic models regardless of data sets, and correlations for the raw HCP data can reach 0.258 and those for the ABCD challenge range from 0.012 to 0.142. But correlations for HCP dramatically drop after adjusting the fluid intelligence score with the method used in the ABCD challenge.

These findings are very interesting. Some possible explanations include: (1) The adolescent's brain is still under development and will have a much bigger variation in reflecting intelligence than adults. This is why in general we see lower prediction power of ABCD compared with HCP with adjusted fluid intelligence scores. (2) Some covariates we regress out from the raw fluid intelligence score are not orthogonal to the T1w imaging data. For example, the brain volume is also a feature of T1w imaging data and can explain part of the fluid intelligence. This finding indicates that the prediction power of brain structure to fluid intelligence we found in the ABCD challenge is very conservative. Brain structure can explain much more variance of fluid intelligence.

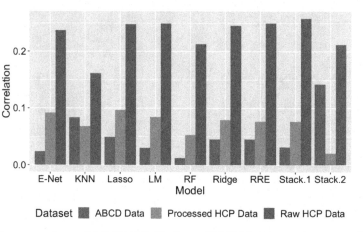

Fig. 4. The comparison of ABCD-NP-Challenge VS HCP data set in terms of correlations between measured fluid intelligence scores and predicted ones. Stack.1 represents the base learners LM and Lasso with the top learner Ridge. Stack.2 represents the base learners KNN and E-Net with the top learner Ridge.

4   Discussion

After comparing the results of ABCD-NP-Challenge and HCP data sets, we can see that ensemble learning, especially the stacking combination method, improves the results of basic models. In terms of prediction, we can combine different base models together to improve the final results. Also, we only rely on the volumetric data extracted from T1w imaging in our modeling. Other features derived from T1w images, such as cortical folding patterns and curvature, might also explain part of the intelligence. Due to time and computation limitation, we do not have the chance to explore these features.

In the HCP, the dramatic drop of prediction power after adjusting the fluid intelligence score tells us that this pre-processing procedure throws away part of the information that relates to brain structure. In other words, the gender, age, race/ethnicity, family factors, and brain volume are not independent with brain structure. We also observe that the fluid intelligence score can be better predicted with the HCP data in general. It might because the brains of children are still under development. The volume of their brains may change day by day and the neural network connections have not been built completely, so it is difficult to predict the fluid intelligence with such dynamic data. Conversely, the brains of young adults have developed to a relatively stable stage. The volume of brain and neural network won't vary too much so that their relationship with intelligence may be easier to capture. But we also want to acknowledge that the two fluid intelligence scores (in ABCD and HCP) are measured differently and the T1w images are measured with different scanners and resolutions and pre-processed with different methods, which might cause the different results.

References

1. Rohlfing, T., Zahr, N.M., Sullivan, E.V., Pfefferbaum, A.: The SRI24 multichannel atlas of normal adult human brain structure. Hum. Brain Mapp. **31**(5), 798–819 (2010)
2. Freund, Y., Schapire, R.E.: A decision-theoretic generalization of on-line learning and an application to boosting. J. Comput. Syst. Sci. **55**(1), 119–139 (1997)
3. Rohlfing, T., Cummins, K., Henthorn, T., Chu, W., Nichols, B.N.: N-CANDA data integration: anatomy of an asynchronous infrastructure for multi-site, multi-instrument longitudinal data capture. J. Am. Med. Inf. Assoc. **21**(4), 758–762 (2013)
4. Dale, A.M., Fischl, B., Sereno, M.I.: Cortical surface-based analysis: I. Segmentation and surface reconstruction. Neuroimage **9**(2), 179–194 (1999)
5. Bilker, W.B., Hansen, J.A., Brensinger, C.M., Richard, J., Gur, R.E., Gur, R.C.: Development of abbreviated nine-item forms of the Raven's standard progressive matrices test. Assessment **19**(3), 354–369 (2012)
6. Zhou, Z.-H.: Ensemble Methods: Foundations and Algorithms, 1st edn. Chapman and Hall/CRC, New York (2012)
7. Breiman, L.: Bagging predictors. Mach. Learn. **24**(2), 123–140 (1996)
8. Friedman, J.: Greedy function approximation: a gradient boosting machine. Ann. Stat. **29**(5), 1189–1232 (2001)
9. Chen, T., Guestrin, C.: XGBoost: a scalable tree boosting system. In: Proceedings of the 22nd ACM SIGKDD International Conference on Knowledge Discovery and Data Mining, pp. 785–794. ACM, New York (2016)
10. Dietterich, T.G.: Ensemble methods in machine learning. In: Kittler, J., Roli, F. (eds.) MCS 2000. LNCS, vol. 1857, pp. 1–15. Springer, Heidelberg (2000). https://doi.org/10.1007/3-540-45014-9_1
11. Cook, R.D., Forzani, L., Zhang, X.: Envelopes and reduced-rank regression. Biometrika **102**(2), 439–456 (2015)

Predicting Fluid Intelligence in Adolescent Brain MRI Data: An Ensemble Approach

Shikhar Srivastava(ID), Fabian Eitel(ID), and Kerstin Ritter$^{(\boxtimes)}$(ID)

Charité — Universitätsmedizin Berlin, Corporate Member of Freie Universität Berlin, Humboldt-Universität zu Berlin, and Berlin Institute of Health (BIH), Department of Psychiatry and Psychotherapy, Bernstein Center for Computational Neuroscience, Berlin Center for Advanced Neuroimaging, 10117 Berlin, Germany
kerstin.ritter@charite.de

Abstract. Decoding fluid intelligence from brain MRI data in adolescents is a highly challenging task. In this study, we took part in the ABCD Neurocognitive Prediction (NP) Challenge 2019, in which a large set of T1-weighted magnetic resonance imaging (MRI) data and pre-residualized fluid intelligence scores (corrected for brain volume, data collection site and sociodemographic variables) of children between 9–11 years were provided ($N = 3739$ for training, $N = 415$ for validation and $N = 4516$ for testing). We propose here the Caruana Ensemble Search method to choose best performing models over a large and diverse set of candidate models. These candidate models include convolutional neural networks (CNNs) applied to brain areas considered to be relevant in fluid intelligence (e.g. frontal and parietal areas) and high-performing standard machine learning methods (namely support vector regression, random forests, gradient boosting and XGBoost) applied to region-based scores including volume, mean intensity and count of gray matter voxels. To further create diversity and increase robustness, a wide set of hyperparameter configurations for each of the models was used. On the validation and the hold out test data, we obtained a mean squared error (MSE) of 71.15 and 93.68 respectively (rank 12 out of 24, MSE range 92.13–102.25). Among most selected models were XGBoost together with the three region-based scores, the other regression models together with volume or CNNs based on the middle frontal gyrus. We discuss these results in light of previous research findings on fluid intelligence.

Keywords: Fluid intelligence · Ensembles · Deep learning · Convolutional neural networks · Structural MRI

1 Introduction

The concept of fluid and crystallized intelligence as two distinct types of general intelligence has been first described by Raymond Cattell [1]. Whereas crystallized intelligence mainly involves knowledge-based reasoning and depends on

© Springer Nature Switzerland AG 2019
K. M. Pohl et al. (Eds.): ABCD-NP 2019, LNCS 11791, pp. 74–82, 2019.
https://doi.org/10.1007/978-3-030-31901-4_9

the level of experience, fluid intelligence defines the ability to flexibly adapt to new problems and solve them logically independent of previous knowledge or past experience [2]. Especially in developing brains of children, fluid intelligence is an essential building block which not only facilitates acquiring new abilities [3,4] but also has been found to accurately predict performance in cognitively demanding tasks at later stages of life [5]. In neuroimaging studies, fluid intelligence has been mostly associated with a large network of parieto-frontal regions including temporal and cingulate cortices (known as parieto-frontal integration theory [6]). A meta-analysis of voxel-based morphometry studies revealed that amounts of gray matter in frontal and temporal cortices as well as subcortical areas are related to intelligence [7].

The ABCD Neurocognitive Prediction Challenge (ABCD-NP-Challenge 2019)[1] invited research teams from around the world to predict pre-residualized fluid intelligence scores based on structural MRI data obtained from the Adolescent Behavioural Cognitive Development (ABCD) study [8]. The ABCD study is the largest long-term study of brain development and child health in the United States, which endeavors to understand how childhood experiences (such as sports or video games) interact with each other and affect brain development [9].

To build a robust and reliable model for predicting fluid intelligence in the ABCD-NP-Challenge 2019, we have chosen an ensemble model searching over a library of models built on deep learning and standard machine learning models. As deep learning method, we have implemented region-based convolutional neural networks (CNNs) which have gained popularity in various paradigms of medical image analysis [10]. Regions have been chosen with respect to the parieto-frontal integration theory. Among standard machine learning methods we used support vector regression (SVR), random forests (RF) and gradient boosting (GB) along with its XGBoost based implementation, which were applied to various region-based summary measures including volume, mean intensity and count of gray matter voxels. As ensemble method, we employed the Caruana Ensemble Search (CES) method [11]. Ensemble approaches generally minimize the risk of overfitting due to suboptimal parameter choices or bias in the data and have been shown to outperform single machine learning models [12,13]. Since they thrive on diversity of models in the library [14,15], we increased diversity by building the models on different input sets (minimally processed MRI volumes and summary measures) in conjunction with diverse supervised learning algorithms (CNNs, SVR, RF, GB and XGBoost) and a wide range of hyperparameter combinations for each algorithm.

2   Methods

2.1   Data Set

The ABCD-NP-Challenge 2019 released structural MRI data in combination with pre-residualized fluid intelligence scores of $N = 8670$ children between 9–11 years from the ABCD study [8]. Structural MRI data included skull stripped

---

[1] Official website: https://sibis.sri.com/abcd-np-challenge/.

T1-weighted (T1w) MRI scans non-linearly aligned to the SRI 24 atlas, gray matter segmentation into regions of interests (ROIs) and corresponding volume scores calculated by [16]. Quality of the pre-processed MRI data were ensured by the challenge organizers.[2] Fluid intelligence scores were measured using the NIH Toolbox Neurocognition battery [17] and pre-residualized on data collection site, brain volume and sociodemographic variables (age, sex, race/ethnicity, highest parental education, parental income and parental marital status) using linear regression. The challenge task was to predict pre-residualized fluid intelligence scores based on the MRI data for individual children. The total data set was split into a training data set ($N = 3739$), a validation set ($N = 415$) and a test set ($N = 4516$). Only for training data and internal validation the pre-residualized fluid intelligence scores were provided in addition to the structural MRI data.

2.2   Data Split and Resampling

We have split the training data into a pure training data set (80%), an ensemble optimization set (15%) and an internal validation set (5%) stratified on pre-residualized fluid intelligence, age and sex. Missing ROI values (less than 0.05%) were imputed with mean values of the same region in the remaining subjects in the training data. As expected, the residualized fluid intelligence scores do not follow a uniform distribution (see Fig. 1). In such cases of imbalanced or skewed target data, prediction algorithms show difficulty in generalizing for subjects away from the mean [18]. Given that ensemble learning and resampling help in resolving this problem [19–21], we performed oversampling for the children at the tail ends and undersampling for the children at the center of the fluid intelligence distribution. The resampled training set was held constant across all models.

2.3   Modeling

For learning pre-residualized fluid intelligence scores from structural MRI data, we propose here an ensemble approach consisting of a large and diverse set of competing models and the Caruana Ensemble Search (CES) for searching the best models among these 'candidate' models (please see Fig. 2 for an overview). In the following we describe the different types of candidate models and the CES method. The full code is provided at https://github.com/shikharsgit/ABCD_NP_Challenge/.

CNNs on T1-Weighted MRI Data: To save computational costs, cancel noise by including prior knowledge and create diversity, we have built separate convolutional neural network (CNN) candidates on a subset of anatomical brain regions defined by the LPBA40 atlas [22]. The subset was chosen in accordance with previous studies on fluid intelligence [6,7,23] and mostly overlaps with the

---

[2] More details on acquisition and pre-processing are provided on the challenge website.

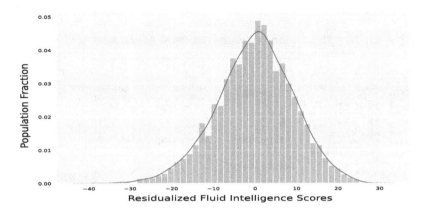

Fig. 1. Density plot of residualized fluid intelligence scores.

widely accepted parieto-frontal integration theory as suggested by Jung and Haier [6]. In particular, we extracted gyrus rectus, hippocampus, inferior frontal gyrus, middle frontal gyrus, postcentral gyrus, precentral gyrus, precuneus, superior frontal gyrus and supramarginal gyrus from T1-weighted MRI data. Thus, in total we created 9 input data sets for each subject, which were min-max scaled to the range $[0, 1]$.

Regarding the configuration of the CNN models, we employed residual connections [24] with up to three types of blocks with each type consisting of two 3D convolutional layers with a varying number of filters (10, 16 or 32) and a rectified linear unit (ReLu) activation. Each convolutional layer was batch normalized and had a drop out rate of 20%. Max pooling was performed before and after the block connections. To create further diversity we built multiple models varying in number of blocks, batch sizes and learning rates. Additionally, we saved training checkpoints at each epoch for the first 15 epochs. In total, 540 CNN candidates were generated. Please see the code at https://github.com/shikharsgit/ABCD_NP_Challenge/ for full details on the different parameter combinations.

SVR, RF, GB and XGBoost on ROI Data: Classical machine learning analyses including support vector regression (SVR), random forests (RF) and gradient boosting (GB) were applied to three types of region-based input data: Mean intensity, number of gray matter voxels (referred to as 'count') and volume. The first two values were calculated for each subject by overlaying the parcellated image over T1-weighted MRI data and extracting either mean intensity or the number of gray matter voxels. Volume was provided by the challenge organizers. Additionally, an optimized version of GB (XGBoost) was applied to the combination of these three measures (referred to as 'all combined').

Similar to the CNN candidates, all features were min-max scaled to the range $[0, 1]$ and we generated a set of candidates for each algorithm by varying the hyperparameters. For SVR, we varied the kernel, the kernel coefficient, the

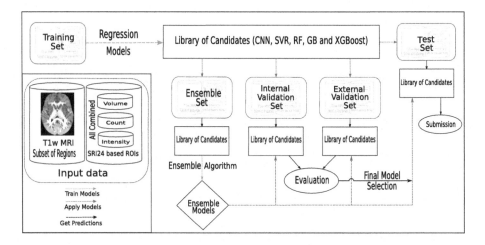

Fig. 2. Modeling framework and evaluation process.

penalty parameter and epsilon [25], resulting in 588 SVR candidates. For both RF and GB, we varied the maximum depth of trees, minimum samples for splitting as well as the number of trees and features [26,27], resulting in 900 RF and 900 GB candidates. For XGBoost, we varied the maximum tree depth, learning rate, number of trees, base learners and sampling ratios of the input data at each tree, each level or each node [28], resulting in 3000 XGBoost candidates. Please see the code at https://github.com/shikharsgit/ABCD_NP_Challenge/ for full details on the different parameter combinations.

Caruana Ensemble Search (CES) Method: CES is a greedy hill-climbing selection method that has been first proposed by Caruana et al. [11] and has been shown to outperform other ensemble methods [29,30]. It produces a weighted linear combination of a subset of candidates by successively adding those models from a library that improve the respective performance measure by some pre-defined threshold on a validation set specified for ensemble optimization. However, when libraries consist of thousands of candidates, there is an increased risk of overfitting. As suggested in [11], we addressed the risk of overfitting in the following ways: First, the ensemble is initialized with the top performing candidates on the ensemble optimization set. Second, variance is reduced by optimizing the ensemble over multiple bags of randomly selected candidates (similar to the bagging method in RF). And third, selection is performed with replacement where models that were added multiple times are averaged according to their weights. Here, we used the CES method to find a set of models from the above described candidate models that result in a preferably low mean squared error (MSE). To save computational time and costs, we pruned our library of candidates by deleting candidates having an MSE of more than 77 and candidates with a Pearson correlation of more than 0.95 with other candidates. By this, we reduced the number of 5928 candidates to 1416 candidates. The CES method was optimized over the ensemble optimization set (15% of training data) and parameters (threshold value and fraction of randomly selected models in each bag) were tuned via a grid-search. As final model, we selected the model which achieved the smallest sum of MSE over the internal (5% of training data) and external validation set ($N = 415$).

3   Results

In Table 1, we present the MSE of the selected CES model in comparison with the top candidates for each algorithm, separately for the ensemble optimization set (15% of training data), the internal validation set (5% of training data) and the external validation set ($N = 415$). The final CES model had a threshold level of 0.001 and 50 bags with 30% of randomly selected models in each bag. The selection of top candidates is based on the performance on the ensemble set. Since the CES model was optimized for the ensemble set, we ignored here the MSE. Best scores are highlighted in bold font. The CES model shows a marginally lower MSE on the external validation set, and the 3rd lowest MSE for the internal validation set.

In Table 2, we represent the top 10 out of a total of 57 selected candidates for the CES model. Most often the XGBoost algorithm with all combined ROI features (i.e. intensity, count and volume) were chosen followed by volume together with either SVR, RF or GB. The most successful CNN model was based on the middle frontal gyrus. Since a large selection of candidates points towards a more robust and stable ensemble as opposed to a repeated selection of a very few candidates, we decided to use the CES model for the final submission in the challenge. Upon declaration of results, our submission on the hold out test data set showed an MSE of 93.68.

Table 1. MSE for the final CES model in comparison with top performing single candidates, separately for the ensemble optimization set, internal validation set and external validation set.

Algorithm	Input data	Ensemble set	Internal val set	External val set
CES	Library of candidates	–	80.78	**71.15**
CNN	Hippocampus	**88.31**	83.39	71.54
SVR	Intensity	89.71	81.60	71.36
RF	Volume	89.59	79.64	71.21
GB	Count	90.10	81.04	71.97
XGBoost	All combined	89.30	**79.21**	72.32

Table 2. Top 10 candidates selected in the CES model ranked by the weights.

Candidate	Input data	Weights
XGBoost	All combined	16.57%
XGBoost	All combined	9.60%
CNN	Middle frontal gyrus	5.62%
SVR	Volume	5.42%
XGBoost	All combined	4.03%
SVR	Volume	3.73%
XGBoost	All combined	3.68%
GB	Volume	3.58%
RF	Volume	3.08%
XGBoost	All combined	2.59%

4   Discussion

Current state-of-the-art prediction algorithms are very powerful in their ability to find non-linear patterns in any kind of data. Especially in medical data sets where individual differences are high and influenced by many factors, those prediction algorithms are usually predisposed to overfit. For the ABCD-NP Challenge 2019, we therefore aimed to build a highly stable model by using a meta-learning ensemble approach searching across a wide range of top-performing deep learning and classical machine learning models. These models were either built directly on T1-weighted MRI data of pre-selected regions or ROI-based data. Notably, none of these models achieved a good performance. Also, the ensemble search method (CES) as a weighted average of 57 underlying models has only marginally improved the validation error (71.15). As evident from the final leader board of this challenge, where all teams submitted their predictions on the hold out test data set, the prediction task itself was very difficult. Our final submission ranked 12th (MSE of 93.68) out of 24 valid submissions (MSE range 92.13–102.25).

Nevertheless, we would like to highlight our region-based CNN modeling approach allowing to incorporate domain knowledge. Even though there was no considerable improvement in prediction from this approach, we note that the CNN models based on the hippocampus (best performing single CNN model) and the middle frontal gyrus (top 3 candidate model in the CES model) contributed to some degree in intelligence prediction. Only a few studies so far have attempted to study neural mechanisms underlying intelligence in children [7]. Shaw et al. [31] have reported that the trajectory of cortical thickness over time is the best predictor of intelligence in developing brains, especially in frontal areas. The ABCD study is an ideal experimental framework for validating those results. In a future study, we would like to extrapolate our analysis with other parts of the brain and also focus on models that are more transparent than ensemble models.

In conclusion, we appreciate the challenge organizers for bringing forth this complex task of fluid intelligence prediction. Such machine learning competitions in combination with large data sets have a high potential to extend research boundaries and to set new benchmarks. Additionally, the chances of overfitting are greatly reduced as the results are evaluated on completely independent test data.

Acknowledgements. We acknowledge support from the German Research Foundation (DFG, 389563835), the Brain & Behavior Research Foundation (NARSAD Young Investigator Grant), the Manfred and Ursula-Müller Stiftung and Charité – Universitätsmedizin Berlin (Rahel-Hirsch scholarship).

References

1. Cattell, R.B.: Intelligence: Its Structure, Growth and Action, vol. 35. Elsevier (1987). https://psycnet.apa.org/record/1987-98151-000
2. Jaeggi, S.M., Buschkuehl, M., Jonides, J., Perrig, W.J.: Improving fluid intelligence with training on working memory. Proc. Natl. Acad. Sci. **105**(19), 6829–6833 (2008). https://doi.org/10.1073/pnas.0801268105
3. Ferrer, E., O'Hare, E.D., Bunge, S.A.: Fluid reasoning and the developing brain. Front. Neurosci. **3**, 3 (2009). https://doi.org/10.3389/neuro.01.003.2009
4. Goswami, U.: Analogical Reasoning in Children. Psychology Press (2013). https://doi.org/10.4324/9781315804729
5. Gottfredson, L.S.: Why g matters: the complexity of everyday life. Intelligence **24**(1), 79–132 (1997). https://doi.org/10.1016/S0160-2896(97)90014-3
6. Jung, R.E., Haier, R.J.: The Parieto-Frontal Integration Theory (P-FIT) of intelligence: converging neuroimaging evidence. Behav. Brain Sci. **30**(2), 135–154 (2007). https://doi.org/10.1017/S0140525X07001185
7. Basten, U., Hilger, K., Fiebach, C.J.: Where smart brains are different: a quantitative meta-analysis of functional and structural brain imaging studies on intelligence. Intelligence **51**, 10–27 (2015). https://doi.org/10.1016/j.intell.2015.04.009
8. Casey, B.J., et al.: The adolescent brain cognitive development (ABCD) study: imaging acquisition across 21 sites. Dev. Cogn. Neurosci. **32**, 43–54 (2018). https://doi.org/10.1016/j.dcn.2018.03.001
9. Adolescent Brain Cognitive Development (ABCD) Study. https://abcdstudy.org/about/
10. Litjens, G., et al.: A survey on deep learning in medical image analysis. Med. Image Anal. **42**, 60–88 (2017). https://doi.org/10.1016/j.media.2017.07.005
11. Caruana, R., Munson, A., Niculescu-Mizil, A.: Getting the most out of ensemble selection. In: Sixth International Conference on Data Mining (ICDM 2006), pp. 828–833. IEEE (2006). https://doi.org/10.1109/ICDM.2006.76
12. Kamnitsas, K., et al.: Ensembles of multiple models and architectures for robust brain tumour segmentation. CoRR, abs/1711.01468 (2017). http://arxiv.org/abs/1711.01468
13. Zhou, Z.-H.: Ensemble Methods: Foundations and Algorithms, 1st edn. Chapman & Hall/CRC (2012). https://doi.org/10.1201/b12207. ISBN 1439830037, 9781439830031
14. Kuncheva, L.I., Whitaker, C.J.: Measures of diversity in classifier ensembles and their relationship with the ensemble accuracy. Mach. Learn. **51**(2), 181–207 (2003). https://doi.org/10.1023/A:1022859003006

15. Sollich, P., Krogh, A.: Learning with ensembles: how overfitting can be useful. In: Advances in Neural Information Processing Systems, pp. 190–196 (1996). http://papers.nips.cc/paper/1044-learning-with-ensembles-how-overfitting-can-be-useful.pdf

16. Pfefferbaum, A., et al.: Altered brain developmental trajectories in adolescents after initiating drinking. Am. J. Psychiatry **175**(4), 370–380 (2018). https://doi.org/10.1176/appi.ajp.2017.17040469. PMID: 29084454

17. Akshoomoff, N., et al.: VIII. NIH toolbox cognition battery (CB): composite scores of crystallized, fluid, and overall cognition. Monogr. Soc. Res. Child Dev. **78**(4), 119–132 (2013). https://doi.org/10.1111/mono.12038

18. Krawczyk, B.: Learning from imbalanced data: open challenges and future directions. Progress Artif. Intell. **5**(4), 221–232 (2016). https://doi.org/10.1007/s13748-016-0094-0

19. Błaszczyński, J., Stefanowski, J.: Neighbourhood sampling in bagging for imbalanced data. Neurocomputing **150**, 529–542 (2015). https://doi.org/10.1016/j.neucom.2014.07.064

20. Galar, M., Fernandez, A., Barrenechea, E., Bustince, H., Herrera, F.: A review on ensembles for the class imbalance problem: bagging-, boosting-, and hybrid-based approaches. IEEE Tran. Syst. Man Cybern. Part C (Appl. Rev.) **42**(4), 463–484 (2012). https://doi.org/10.1109/TSMCC.2011.2161285

21. Krawczyk, B., Woźniak, M., Schaefer, G.: Cost-sensitive decision tree ensembles for effective imbalanced classification. Appl. Soft Comput. **14**, 554–562 (2014). https://doi.org/10.1016/j.asoc.2013.08.014

22. Shattuck, D.W., et al.: Construction of a 3D probabilistic atlas of human cortical structures. Neuroimage **39**(3), 1064–1080 (2008). https://doi.org/10.1016/j.neuroimage.2007.09.031

23. Colom, R., et al.: Hippocampal structure and human cognition: key role of spatial processing and evidence supporting the efficiency hypothesis in females. Intelligence **41**(2), 129–140 (2013). https://doi.org/10.1016/j.neuroimage.2007.09.031

24. He, K., Zhang, X., Ren, S., Sun, J.: Deep residual learning for image recognition. In: Proceedings of the IEEE Conference on Computer Vision and Pattern Recognition, pp. 770–778 (2016). https://doi.org/10.1109/CVPR.2016.90

25. Chang, C.-C., Lin, C.-J.: LIBSVM: a library for support vector machines. ACM Trans. Intell. Syst. Technol. (TIST) **2**(3), 27 (2011). https://doi.org/10.1145/1961189.1961199

26. Breiman, L.: Random forests. Mach. Learn. **45**(1), 5–32 (2001). https://doi.org/10.1023/A:1010933404324

27. Friedman, J.H.: Greedy function approximation: a gradient boosting machine. Ann. Stat. 1189–1232 (2001). https://www.jstor.org/stable/2699986

28. Chen, T., Guestrin, C.: XGBoost: a scalable tree boosting system. In: Proceedings of the 22nd ACM SIGKDD International Conference on Knowledge Discovery and Data Mining, pp. 785–794. ACM (2016). https://doi.org/10.1145/2939672.2939785

29. Whalen, S., Pandey, G.: A comparative analysis of ensemble classifiers: case studies in genomics. In: 2013 IEEE 13th International Conference on Data Mining, pp. 807–816. IEEE (2013). https://doi.org/10.1109/ICDM.2013.21

30. Lessmann, S., Baesens, B., Mues, C., Pietsch, S.: Benchmarking classification models for software defect prediction: a proposed framework and novel findings. IEEE Trans. Softw. Eng. **34**(4), 485–496 (2008). https://doi.org/10.1109/TSE.2008.35

31. Shaw, P., et al.: Intellectual ability and cortical development in children and adolescents. Nature **440**(7084), 676 (2006). https://doi.org/10.1038/nature04513

Predicting Fluid Intelligence from Structural MRI Using Random Forest regression

Agata Wlaszczyk[1] , Agnieszka Kaminska[2] , Agnieszka Pietraszek[3],
Jakub Dabrowski[4], Mikolaj A. Pawlak[5,6] , and Hanna Nowicka[7(✉)]

[1] Tooploox, Wroclaw, Poland
agata.wlaszczyk@gmail.com
[2] Adam Mickiewicz University in Poznan, Poznan, Poland
kaminska@tuta.io
[3] Lodz, Poland
agnieszka.pietraszek86@gmail.com
[4] Poznan University of Technology, Poznan, Poland
jakub.dabrowski.poland@gmail.com
[5] Poznan University of Medical Sciences, Poznan, Poland
[6] Inteneural Networks Inc., Chicago, USA
mpawlak@ump.edu.pl
[7] FMRIB, Wellcome Centre for Integrative Neuroimaging,
University of Oxford, Oxford, UK
hanna.nowicka@keble.ox.ac.uk

Abstract. Fluid intelligence (FI) indicates a set of general abilities like pattern recognition, abstract thinking, and problem-solving. FI is related to inherent, biological factors. We present a method to predict the fluid intelligence score in children (9–10 y/o) from their structural brain scans. For the purposes of this work, we used features derived from the T1-weighted Magnetic Resonance scans from the ABCD study. We used data from 3739 subjects for training and 415 for validation of the model. As features we used the volumes of gray matter regions of interest provided by the challenge organizers, as well as three additional groups of features. These include signal intensity features based on the ROIs, as well as shape-based features derived from the anterior and posterior cross sectional area of the corpus callosum. We used the random forest regressor model for prediction. We compare its performance to other regression-based models (XGBoost Regression and Support Vector Regression). Additionally, we ran a mean decrease accuracy (MDA) algorithm to select features that had high influence on the prediction results. The results we have obtained for the validation set were as follows: MSE = 67.39, R-squared = 0.0762. The proposed method showed promising results and has the potential to provide a good prediction of fluid intelligence based on structural brain scans.

Keywords: Adolescence development · Fluid intelligence · Random Forest Regressor

Majority of the work was done at the BrainHack Warsaw hackathon.

© Springer Nature Switzerland AG 2019
K. M. Pohl et al. (Eds.): ABCD-NP 2019, LNCS 11791, pp. 83–91, 2019.
https://doi.org/10.1007/978-3-030-31901-4_10

1   Introduction

Fluid intelligence (FI) is derived from human cognitive abilities theory where general intelligence is differentiated by its developmental origin into two factors: crystallized and fluid intelligence [1]. The former includes acquired intellectual skill set connected with education and socioeconomic background [2–4], while FI describes general abilities like pattern recognition, abstract thinking, and problem-solving. After decades of misconception around genetic and brain-related mechanisms underlying certain levels of cognitive abilities [5], recent genome-wide association evidence shows that 50% of the inherited genome sequence differences account for heritability of intelligence and that 20% out of this gene pool has been identified [6].

General intelligence is strongly linked to whole brain volume, which is the largest predictor of FI variance in adults (around 12%) when compared to other features such as cortical thickness, white matter structure, white matter hyperintensity load, iron deposits, and microbleeds [7]. In combination, these six variables account for approximately 18–21% of the variance in measured intelligence. Nevertheless, it needs to be noted that this rate was obtained for a general intelligence factor, and for FI the results were less conclusive.

There are several brain regions that can be linked with fluid intelligence. One of them is the prefrontal cortex, which is mentioned in the literature to be essential for relevant cognitive performance based on functional connectivity [8], as well as functional [9,10], structural and neuropsychological evidence [7,9].

The existing literature also mentions the characteristics of cortical architecture as a second potential factor of interest. Recent studies have shown a longitudinal link between changes in white matter microstructure and FI-related performance [7], as well as a robust relationship between FI score and WM integrity using diffusion MRI measures in middle-aged individuals [15].

Another region of interest when considering the link between FI and structure of the regional anatomy is the corpus callosum. That relationship, however, is complex and often overshadowed by the impact of the global brain and white matter volumes [12]. This effect is especially pronounced in adolescents where rapid head growth take place earlier in girls than in boys and neuronal pruning is critical for brain maturation. The relative differences in corpus callosum cross-sectional area have been previously associated with intelligence [13], however studies investigating larger samples have not confirmed that effect [14].

The project presented here was created as an entry to the ABCD-NP challenge [27]. The aim was to build a model to predict a fluid intelligence score based on structural images of childrens' (9–10 y/o) brains.

2   Methods

2.1   Data

The dataset provided by the ABCD-NP challenge organizers was a subset of the ABCD study dataset [16,17]. It consisted of 8669 subjects (48% female) between

107 and 133 months old. For each subject, a T1-weighted structural brain MRI scan was provided as data and, for the training and validation subsets of subjects, also their residualized fluid intelligence scores.

FI scores in the ABCD study were assessed using the NIH Toolbox Neurocognition battery [11]. In order to minimise the impact of confounds that are not related to the brain structure, the raw scores were residualized by the challenge organizers on data collection site, sociodemographic and socioeconomic variables, as well as the total brain volume [27]. Although the distribution of labels seemed to follow a normal distribution (with $\mu = 0.05$, $\sigma = 9.27$) visually, our analysis using the omnibus test of normality implemented in the Python Scipy package showed otherwise ($K^2 = 38.5$, $p < 0.05$).

MRI Data. T1-weighted MRI scans provided by the challenge organizers were preprocessed as described on the challenge website [17,27]. In brief, that involved noise removal, field inhomogeneity correction, brain extraction and affine registration to a common space. The challenge organizers also provided a gray matter parcellation into the ROIs defined by the SRI24 atlas and the volumes of these regions [17].

2.2   Features

We decided to expand the range of features beyond the gray matter region volumes provided by the challenge organizers by generating more features describing brain regions and their characteristics that could be relevant to the problem according to the neuroscientific literature.

Intensity Based Features. Due to the critical role of myelination in cortical connectivity, the prediction of FI based on T1-weighted signal without quantitative measures of white matter development is likely to be limited. It was shown in the literature that T1/T2 maps are good predictors of cortical myelination [18,19]. Having only T1-weighted images in our dataset, we referred to the reports from pathological studies that indicated a link between regional water content and the degree of myelination, which indicate that a qualitative measure of myelination can be based on T1-weighted signal alone [20]. To estimate the degree of cortical myelination we computed three measures: entropy, mean and standard deviation of cortical T1 signal, using fslstats from FSL software package [22], for each atlas-based subregion. These three measures together have the capacity to provide insight into the relative myelin content, based on the T1-weighted signal [21]. Even though the computed metric of region-based signal intensity is just a surrogate measure of water content, which decreases with myelination and cortex maturation, we included it in the analysis as patch-based, local features descriptor, as incorporating low-level visual features is commonly used in the machine learning image-related problems [23].

Features Based on Corpus Callosum Assessment. The final set of additional features was based on the quantification of the subsections of the cross-sectional area of the corpus callosum that have been associated with intelligence [14]. Calculations were performed using the yuki 2.1 tool (part of Automatic Registration Toolbox) [24]. For each subject we computed the total cross-sectional area, length, and perimeter of subregions, based on the Hampel and Witelson divisions [25].

Final Dataset Details. We used data from 3739 subjects for training and 415 for validation of the model. The testing set, whose FI scores were unseen, consisted of 4402 subjects. For each subject we used the volumes of gray matter regions of interest (ROI), signal-based features and corpus callosum properties. Each feature was z-score normalized.

2.3   Model

The approach adopted for this challenge was focused on creating an accurate statistical model of FI. We investigated the following machine learning models: Support Vector Regression (SVR), Random Forest Regression, and XGBoost Regression (XGB). SVR works by determining a hyperplane with a proper margin of tolerance. In a Random Forest, a number of decision trees are fitted based on randomly chosen subsamples from the dataset, and the results from individual trees are pooled to prevent overfitting and to get the best possible predictive accuracy. In our experiments we set up a fixed numbers of 500 trees. The XGB model, like the Random Forest, makes use of an ensemble of weak learners and combines them to boost the model's performance. All of the models and statistical methods used for data analysis were implemented in Python.

One of the methods for increasing the quality of predictions was performing feature selection. The feature groups were defined with respect to their origin: (a) the default features provided by the organizers and corresponding to ROI volumes, (b) signal-based features, (c) corpus callosum properties. To test the importance of the feature groups, each possible combination of these groups was fed into each of the selected models as the training data and evaluated with mean squared error (MSE) on the validation set. Additionally, we implemented a feature importance evaluation step to eliminate individual features that would potentially introduce noise or decrease the model's metrics. The method we used is known as mean decrease accuracy [26]. It takes samples from the validation set and iteratively replaces each of its features with a random noise (or shuffled values from a given column) to check how the performance changes when a feature is distorted. The method's output is a weight for each feature that indicates its importance for a model's performance. The higher the weight is, the more important the feature is for the model. All experiments were compared to two baseline methods: a random regressor, which predicts a random value from the Gaussian distribution with the mean and the standard deviation obtained from the training labels ($MSE = 164.69$ on validation subset of the data) and a majority regressor, which always predicts 0 ($MSE = 71.8$ on validation data). During the experiments, no cross-validation was performed and the entire validation set was used to evaluate the model's quality.

3   Results

Tables 1 and 2 show the performance of the considered models and feature groups in terms of MSE on the validation set. Each row indicates a separate combination of various feature groups, where group (a) corresponds to the features provided by the organizers, group (b) relates to the image-based features and group (c) consists of features derived from the corpus callosum. Table 1 demonstrates the results without the feature selection step, and Table 2 shows the results augmented with feature importance evaluation and selection. In the current experiment, the threshold for distinguishing relevant features from irrelevant features was chosen arbitrarily to correspond to the 75th percentile of the weights distribution for each combination of model and features. We only took the features with the highest weights, as lower and negative weights contribute to decreasing the model's performance. However, repeated experiments with various threshold values could lead to a more consistent overview of how the results change given the varying number of relevant features.

Table 1. MSE results of all considered models for all combinations of feature groups without feature selection step. The feature groups (a), (b) and (c) refer to the features provided by the organizers, image-based features and corpus callosum properties respectively.

	SVR	RF	XGB
(a)	73.32	71.17	74.68
(b)	72.17	71.56	76.46
(c)	72.5	73.1	76.68
(a,b)	72.35	71.53	79.58
(a,c)	72.8	70.98	73.89
(b,c)	71.95	71.28	78.7
(a,b,c)	72.18	71.23	77.45

For the final experiment, we chose the Random Forest Regressor, after comparing it with other models using the same set of features (see Table 2). The final score achieved on the validation set for the Random Forests with additional feature importance evaluation step was $MSE = 67.39$ with $R^2 = 0.0762$. The error on the test set was $MSE = 92.93$, which corresponds to the 6th place on the final contest leader board.

Table 2. MSE results of all considered models for all combinations of feature groups with the feature selection step. The feature groups are defined as in Table 1.

	SVR	RF	XGB
(a)	69.24	68.58	74.62
(b)	72.64	71.56	77.12
(c)	73.48	77.02	78.37
(a,b)	68.45	68.75	76.77
(a,c)	69.79	68.37	75.43
(b,c)	73.54	71.72	79.23
(a,b,c)	69.99	**67.39**	68.68

3.1   Analysis of Selected Features

We observed the weights resulting from the feature selection step to fall in the range between −0.19 and 0.15. For the final training, out of the initial 541 features, we used only 141, which passed the 75th percentile threshold. Among those 141 selected 48 came from the default set of features, 9 were selected from the corpus callosum features and 84 originated from the low-level intensity-based features. The fraction of the feature subset that survived thresholding (percentage of the selected features with respect to all features originating from a given group) was 39.3%, 56.3% and 20.9% respectively.

4   Discussion

We present a method for predicting fluid intelligence scores from T1-weighted MRI scans of childrens' brains. The proposed model has shown promising results with MSE value of 67.39 on the validation data, which was the best result out of the submissions to the challenge based on the validation leaderboard. However, we observed that our mean squared error is only 6.5% smaller than what could be obtained by just feeding zeroes as the prediction. That indicates that there is still scope for improvement in terms of the accuracy and generalizability of the model.

Analysis of the scores (especially low R^2) shows that we still lack sufficiently discriminative features to account for the majority of the variation in the residualized intelligence scores. To gain a better understanding of the results generated by our model, we compared the distributions of the intelligence scores from the training set and the distributions predicted by our model on the test set. One of the most important problems noticed during the evaluation of predictions was a divergence of their distribution from the initial distribution of training labels. The training set scores fell within a range from −38.8 to 28.1, whereas the distributions of the predictions of the test set were from −5.1 to 3.5. In particular, even though the means of the mentioned distributions were close to each other (0.003 for the generated test predictions in comparison to 0.05 for the training labels), the difference of standard deviations of the mentioned groups was far larger (1.24 for the predictions, as opposed to 9.27 for the training labels). This may indicate that either the collected features did not manage to capture the characteristics of subjects who exhibit large deviation from the mean or that the model was unable to learn these distinctions. We identify this observation as a potential deficiency in the proposed solution and as an important issue to tackle in future experiments.

The increase in the error observed when comparing the model's performance on the validation and the test sets suggests that the model overfits to the available data. It is also possible that the FI scores in the testing set had a different distribution than that of the validation set, given that for most of the teams the MSE on the testing set increased substantially. Current results and approach taken in this study lean towards the view that fluid intelligence is a much more complex phenomenon that cannot be satisfactory captured with solely structural information of various brain regions.

Future Work Directions. Given the time constraints, we trained the model using only a small portion of the sensible data augmentation options. In future extensions of the work we are planning to expand the feature space with other biologically informed features. One of the areas that is worth considering is focussing on other subcortical regions, such as the basal ganglia [31].

It would also be worth trying other modelling approaches. We expect that there are potential improvements that could be brought by more advanced models, especially based on the neural networks. In particular, recent advances in the field of convolutional neural networks (CNNs) offer promising possibilities [32], as they can be applied both as the final models and as means to retrieve additional deep features. We believe that exploring this technique has a potential, since previous research [28] has proven to yield good results in a task that combined CNNs with a regression problem for brain MRI images. Another idea would be to try a slightly different approach, potentially an ensemble of models. For instance starting from dividing the data into bins of labels ranges and building a model for first classification and then, within the bins, regression. In this way, it might be possible to obtain more fine-grained regression models that may be able to capture subtle differences in narrower distributions (lower variance).

The challenge data was limited to the T1 images. Outside of the challenge we believe that use of T2-weighted or diffusion-weighted images would allow better maps of myelination to be created [29,30].

5   Summary

We presented our approach for predicting the fluid intelligence scores from the T1-weighted structural MRI scans of the brain. After comparing with other regression models we used a Random Forest Regression model, as it showed the best performance. We used selected sets of features based on the neuroscientific literature. We found that not all of the relationships reported from the studies translate to informative features. Despite our model showing relatively good performance compared to other challenge submissions, our solution predicts only 6.5% of the variance indicating scope for improvement, therefore we also propose further possible directions.

Acknowledgement. The authors would like to thank Professor Mark Jenkinson for his helpful comments regarding the manuscript.

References

1. Cattell, R.B.: Abilities: Their Structure, Growth, and Action. Houghton Mifflin, Oxford (1971)
2. Conway, A.R.A., Kovacs, K.: New and emerging models of human intelligence. Wiley Interdisc. Rev.: Cognitive Sci. **6**(5), 419–426 (2015)
3. Kaya, F., Stough, L.M., Juntune, J.: Verbal and nonverbal intelligence scores within the context of poverty. Gifted Educ. Int. **33**(3), 257–272 (2016)
4. Rindermann, H., Flores-Mendoza, C., Mansur-Alves, M.: Reciprocal effects between fluid and crystallized intelligence and their dependence on parents' socioeconomic status and education. Learn. Individ. Differ. **20**(5), 544–548 (2010)
5. Gottfredson, L.S.: Hans Eysenck's theory of intelligence, and what it reveals about him. Personality Individ. Differ. **103**, 116–127 (2016)
6. Plomin, R., Stumm, S.V.: The new genetics of intelligence. Nat. Rev. Genet. **19**, 148–159 (2018)
7. Ritchie, S.J., et al.: Beyond a bigger brain: multivariable structural brain imaging and intelligence. Intelligence **51**, 47–56 (2015)
8. Cole, M.W., Yarkoni, T., Repovs, G., Anticevic, A., Braver, T.S.: Global connectivity of prefrontal cortex predicts cognitive control and intelligence. J. Neurosci. **32**(26), 8988–8999 (2012)
9. Choi, Y.Y., et al.: Multiple bases of human intelligence revealed by cortical thickness and neural activation. J. Neurosci. **28**(41), 10323–10329 (2008)
10. Gray, J.R., Chabris, C.F., Braver, T.S.: Neural mechanisms of general fluid intelligence. Nat. Neurosci. **6**(3), 316 (2003)
11. Akshoomoff, N., et al.: NIH toolbox cognition battery (cb): Composite scores of crystallized, fluid, and overall cognition. Monogr. Soc. Res. Child Dev. **78**(4), 119–132 (2013)
12. McDaniel, M.A.: Big-brained people are smarter: a meta-analysis of the relationship between in vivo brain volume and intelligence. Intelligence **33**, 337–346 (2005)
13. Luders, E., Thompson, P.M., Narr, K.L., Zamanyan, A., Chou, Y.Y., Gutman, B., Dinov, I.D., Toga, A.W.: The link between callosal thickness and intelligence in healthy children and adolescents. Neuroimage **54**(3), 1823–30 (2011)
14. Westerhausen, R., et al.: The corpus callosum as anatomical marker of intelligence? A critical examination in a large-scale developmental study. Brain Struct Funct. **223**(1), 285–296 (2018)
15. Haász, J., Westlye, E.T., Fjær, S., Espeseth, T., Lundervold, A., Lundervold, A.J.: General fluid-type intelligence is related to indices of white matter structure in middle-aged and old adults. Neuroimage **83**, 372–383 (2013)
16. ABCD study website abcdstudy.org. Accessed 4 Apr 2019
17. Data Supplement of Pfefferbaum et al.: Altered Brain Developmental Trajectories in Adolescents After Initiating Drinking. Am. J. Psychiatry **175**(4), 370–380 (2018)

18. Glasser, M.F., Van Essen, D.C.: Mapping human cortical areas in vivo based on myelin content as revealed by T1- and T2-weighted MRI. J. Neurosci. **31**(32), 11597–11616 (2011)
19. Ganzetti, M., Wenderoth, N., Mantini, D.: Whole brain myelin mapping using T1- and T2-weighted MR imaging data. Front Hum Neurosci. **8**, 671 (2014)
20. Koenig, S.H.: Cholesterol of myelin is the determinant of gray-white contrast in MRI of brain. Magn. Reson. Med. **20**(2), 285–291 (1991)
21. Sigalovsky, I.S., Fischl, B., Melcher, J.R.: Mapping an intrinsic MR property of gray matter in auditory cortex of living humans: a possible marker for primary cortex and hemispheric differences. Neuroimage **32**(4), 1524–37 (2006)
22. Jenkinson, M., Beckmann, C.F., Behrens, T.E., Woolrich, M.W., Smith, S.M.: FSL. NeuroImage **62**, 782–90 (2012)
23. Zhang, X., Yang, Y.H., Han, Z., Wang, H., Gao, C.: Object class detection: a survey. ACM Comput. Surv. (CSUR) **46**(1) (2013)
24. Automatic Registration Toolbox. https://www.nitrc.org/projects/art/. Accessed 4 Apr 2019
25. Ardekani, B.A., Figarsky, K., Sidtis, J.J.: Sexual dimorphism in the human corpus callosum: an MRI study using the OASIS brain database. Cereb. Cortex **23**(10), 2514–2520 (2012)
26. Han, H., Guo, X., Yu, H.: Variable selection using mean decrease accuracy and mean decrease gini based on random forest. In: 2016 7th IEEE International Conference on Software Engineering and Service Science (ICSESS), pp. 219–224. IEEE (2016)
27. Website of the challenge. https://sibis.sri.com/abcd-np-challenge/. Accessed 4 Apr 2019
28. Cole, J.H., et al.: Predicting brain age with deep learning from raw imaging data results in a reliable and heritable biomarker. NeuroImage **163**, 115–124 (2017)
29. Abdollahi, R.O., et al.: Correspondences between retinotopic areas and myelin maps in human visual cortex. Neuroimage **99**, 509–524 (2014)
30. Ganzetti, M., Wenderoth, N., Mantini, D.: Whole brain myelin mapping using T1-and T2-weighted MR imaging data. Frontiers Hum. Neurosci. **8**, 671 (2014)
31. Burgaleta, M., et al.: Subcortical regional morphology correlates with fluid and spatial intelligence. Hum. Brain Mapp. **35**(5), 1957–1968 (2014)
32. Bernal, J., et al.: Deep convolutional neural networks for brain image analysis on magnetic resonance imaging: a review. Artif. Intell. Med. **95**, 64–81 (2018)

Nu Support Vector Machine in Prediction of Fluid Intelligence Using MRI Data

Yanli Zhang-James[1(✉)] ⓘ, Stephen J. Glatt[1,2] ⓘ,
and Stephen V. Faraone[1,3] ⓘ

[1] Department of Psychiatry and Behavioral Sciences,
SUNY Upstate Medical University, Syracuse, NY, USA
zhangy@upstate.edu
[2] Psychiatric Genetic Epidemiology and Neurobiology Laboratory PsychGENe
Lab, SUNY Upstate Medical University, Syracuse, NY, USA
[3] Department of Neuroscience and Physiology,
SUNY Upstate Medical University, Syracuse, NY, USA

Abstract. In response to the ABCD Neurocognitive Prediction Challenge (ABCD-NP-Challenge 2019), we developed machine learning algorithms to predict the fluid intelligence (FI) score using T1-weighed magnetic resonance imaging (MRI) data. 122 volumetric scores of regions of interest from 3739 samples provided in the training set were used to train the models and 415 samples were assigned as validation samples. We performed feature reduction using principal factors factor analysis on the training set volume. 36 factors explaining 100% of the total variance were extracted; the top 18 explained 80% of the variance. We estimated three types of models based on: (1) all regional brain volumes, (2) the 18 top factors or (3) the 36 complete factors. We used Scikit-Learn's grid search method to search the hyperparameter spaces of eight different machine learning algorithms. The best model, a Nu support vector regression model (NuSVR) using 36 factor scores as features, yielded the highest validation score ($R^2 = 0.048$) and a relatively low training score (0.22), the latter of which was important for reducing the degree of over-fitting. The mean squared errors (MSEs) for the training and validation samples were 68.2 and 68.6, respectively; the correlation coefficients were 0.54 and 0.21 ($p < 0.0001$ for both). The final MSE for the test set was 95.63. Learning curve analysis suggests that the current training sample size is still too small; increasing sample size should improve predictive accuracy. Overall, our results suggest that, given a large enough sample, machine learning methods with structural MRI data may be able to accurately estimate fluid intelligence.

Keywords: Machine learning · MRI · Fluid intelligence

1 Introduction

Fluid intelligence (FI) is refers to the ability to reason and to solve new problems independently of previously acquired knowledge [1]. Many studies have examined the correlations of brain anatomical structures with human intelligence [1–3]. Deciphering the neural mechanisms underlying human intelligence is important for understanding

© Springer Nature Switzerland AG 2019
K. M. Pohl et al. (Eds.): ABCD-NP 2019, LNCS 11791, pp. 92–98, 2019.
https://doi.org/10.1007/978-3-030-31901-4_11

neurocognitive development and has fundamental implications for education and clinical work. For example, different brain regions and anatomical differences have been associated with different learning abilities. In children, anatomical differences in hippocampus and associated learning and memory regions predicted difference in math skill acquisition [4]. In adults, video game skills were linked to striatal volumes [5] and foreign language skills were linked to white matters of the left insula/prefrontal cortex and inferior parietal cortices [6]. Better characterization of brain biomarkers for different cognitive functions may facilitate the development of targeted training and intervention programs for both typically developing children and those with developmental disabilities.

As the largest long-term study of brain development and child health in the United States [7], the Adolescent Brain Cognitive Development Study (ABCD) Consortium invited researchers to participate in their Neurocognitive Prediction (NP) Challenge. Contestants were asked to use structural MRI data acquired from the study to predict fluid intelligence scores. For the NP Challenge, data from 4154 subjects were provided to participants for training (3739 samples) and validation (415 samples). Data from 4515 additional subjects was reserved as the test set. This report describes the results of our machine learning analyses of these data.

2 Methods

2.1 Dataset and Features

Fluid intelligence (FI) scores were pre-residualized by the Challenge organizer to remove the effects of total brain volume and sociodemographic variables including data collection site, age at baseline, sex at birth, race/ethnicity, highest parental education, parental income, and parental marital status. FI scores from the training and validation sets were provided to contestant during the challenge. 122 regional brain volumes were provided by the Challenge organizer for all samples from the training, validation and test sets. Briefly, the skull stripped MRI images were affinely aligned to the SRI 24 atlas [9], and segmented into regions of interest according to the atlas using standardized processing pipelines in the ABCD Study (details see [10]).

We factor analyzed the training set, and 36 principal factors were identified to account for 100% of the variances with the top 18 factors accounted for 80% of the total variances. Varimax rotated factor scores were generated for the validation and test sets based on the training set factors. In addition to the MRI volumetric scores and factor scores, we included age and sex as features as they are universally available information. All input features were scaled based on the training set's minimum and maximum values.

2.2 Statistical Machine Learning Methods

Eight different algorithms were investigated using Scikit-Learn's grid search tool for hyper-parameter optimization: random forest regressor (RFR), stochastic gradient descent regressor (SGD), Lasso linear model with least angle regression (LassoLar), elastic net (EN), multilayer perceptron regressor (MLP), Ridge regression (Ridge), support vector regression (SVR) and Nu support vector regression (NuSVR) with linear, poly or sigmoid kernel. See the Scikit-Learn online documentation for detailed hyper-parameters available for tuning using grid search tool (https://scikit-learn.org). During grid search, models were optimized using the R^2 coefficient of the validation samples. R^2 is defined as follows:

$$R^2 = 1 - \frac{u}{v}$$

where u is the residual sum of squares

$$u = \sum_{i=1}^{n} \left(y_{true} - y_{pred} \right)^2$$

and v is the total sum of squares

$$v = \sum_{i=1}^{n} \left(y_{true} - \frac{1}{n} \sum_{i=1}^{n} y_{true} \right)^2$$

A perfect prediction will have a score of 1.

We compared the training and validation R^2 scores for different models using different input features. The best model was chosen based on having the highest validation score and the lowest training score that was as good as or better than the validation score. We used the latter decision rule to reduce overfitting.

Prediction of FI scores was obtained by fitting training data to the best model and obtaining the predicted values (y_{pred}). The predicted results were also evaluated with the mean squared error (MSE, requested by the Challenge organizer, see below) and correlation coefficient with the true FI values (y_{true}):

$$MSE = \frac{1}{n} \sum_{i=1}^{n} \left(y_{pred} - y_{true} \right)^2$$

2.3   Learning Curve Analyses

Learning curves were generated by using deciles of the total training sample to fit the model. For each decile of sample size, we plotted the training and validation R^2. We use learning curve analysis to evaluate the model's bias and variance, as well as the sample size effect. This helps to draw inferences about how models might be improved in the future [11].

3 Results

3.1 Grid Search Results

The grid searches of hyperparameter spaces for eight models using three different feature sets returned 24 sets of training and validation R^2 scores (Fig. 1). Each dot within each of the 24 sets represents one specific model with a unique set of hyperparameters setting. NuSVR had the highest validation scores (the highest $R^2 = 0.046$). RFR ranked second, but was considerable lower (the highest $R^2 = 0.029$). None of the other models yielded predictions with R^2 above 0.02. Thus we only considered the NuSVR model further. For NuSVR, the best feature set used the 36 principal factor scores as inputs. The best hyperparameters were: *gamma* = 1, *nu* = 0.2, C = 10, kernel = 'rbf'.

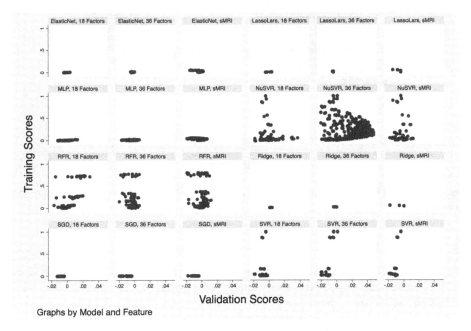

Fig. 1. Training scores were plotted against validation scores (R^2) for all eight models using three types of input features. Validation scores below $-.02$ were discarded. sMRI, structure MRI.

3.2 Prediction Results

Using the best NuSVR model, we predicted the FI scores for the training and the validation sets. Predicted scores were plotted against the actual FI scores (Fig. 2). The scores were significantly correlated ($r = 0.54$ for the training samples and 0.21 for the validation samples ($p < 0.0001$ for both). The mean squared errors (MSE) for the training and validation samples were 68.2 and 68.6, respectively. Removing age and sex as predictors from the model yielded a slightly higher MSE (70.5), but the predictions was significantly different from the original prediction ($F_{(1, 17314)} = 0.72$; $p = 0.4$).

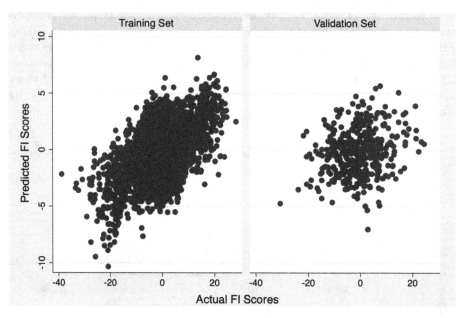

Fig. 2. Predicted Fluid Intelligence (FI) score vs the actual FI score were plotted for training and validation sets.

3.3 Learning Curve

The learning curve plots the training and validation R^2 for 10 incremental sample sizes (at 10% increments) up to the total number of samples. In an ideal model, with increasing sample sizes, the training and validation scores should gradually converge. We found, however, that our training and validation scores did not fully converge, suggesting that model has some degree of over-fitting at the current sample size and that increasing the training sample will be needed to improve the accuracy. Having more predictive features or using current features in a more efficient way could also increase accuracy (Fig. 3).

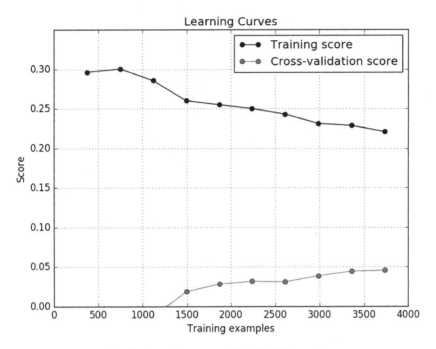

Fig. 3. Learning curve of the NuSVR model.

4 Discussion

Using the structural MRI data provided by the ABCD NP Challenge organizer, we developed statistical machine learning algorithms to predict fluid intelligence scores. We identified NuSVR to be the best prediction model and found that using 36 principal factors yielded the highest prediction accuracies. The predicted FI scores were significantly, albeit modestly, correlated with the actual scores. Our results show the promise of using structural MRI data to predict fluid intelligence and support prior findings of strong anatomical correlations of brain structures with human intelligence. However, we also found that current sample size is not adequate and that more training samples will likely help to improve the model's prediction.

In addition to the sample size limit, we also note that we did not use the T1-weighted MRI images that were also provided for the Challenge. Methods such as convolutional neural networks may be able to extract useful features from the three dimensional MRI images. Such methods may help improve prediction accuracies.

References

1. Jaeggi, S.M., et al.: Improving fluid intelligence with training on working memory. Proc. Natl. Acad. Sci. U.S.A. **105**(19), 6829–6833 (2008)
2. Li, Y., et al.: Brain anatomical network and intelligence. PLoS Comput. Biol. **5**(5), e1000395 (2009)

3. Wang, L., et al.: MRI-based intelligence quotient (IQ) estimation with sparse learning. PLoS ONE **10**(3), e0117295 (2015)
4. Supekar, K., et al.: Neural predictors of individual differences in response to math tutoring in primary-grade school children. Proc. Natl. Acad. Sci. U.S.A. **110**(20), 8230–8235 (2013)
5. Erickson, K.I., et al.: Striatal volume predicts level of video game skill acquisition. Cereb. Cortex **20**(11), 2522–2530 (2010)
6. Golestani, N., Pallier, C.: Anatomical correlates of foreign speech sound production. Cereb. Cortex **17**(4), 929–934 (2007)
7. Feldstein Ewing, S.W., Bjork, J.M., Luciana, M.: Implications of the ABCD study for developmental neuroscience. Dev. Cogn. Neurosci. **32**, 161–164 (2018)
8. Pedregosa, F., et al.: Scikit-learn: machine learning in python. J. Mach. Learn. Res. **12**, 2825–2830 (2012)
9. Rohlfing, T., et al.: The SRI24 multichannel atlas of normal adult human brain structure. Hum. Brain Mapp. **31**(5), 798–819 (2010)
10. Pfefferbaum, A., et al.: Altered brain developmental trajectories in adolescents after initiating drinking. Am. J. Psychiatry **175**(4), 370–380 (2018)
11. Webb, G.I., et al.: Learning Curves in Machine Learning, pp. 577–580 (2011)

An AutoML Approach for the Prediction of Fluid Intelligence from MRI-Derived Features

Sebastian Pölsterl$^{(\boxtimes)}$, Benjamín Gutiérrez-Becker, Ignacio Sarasua, Abhijit Guha Roy, and Christian Wachinger

Artificial Intelligence in Medical Imaging (AI-Med),
Department of Child and Adolescent Psychiatry,
Ludwig Maximilian Universität, Munich, Germany
{sebastian,benjamin,ignacio,abhijit,christian}@ai-med.de

Abstract. We propose an AutoML approach for the prediction of fluid intelligence from T1-weighted magnetic resonance images. We extracted 122 features from MRI scans and employed Sequential Model-based Algorithm Configuration to search for the best prediction pipeline, including the best data pre-processing and regression model. In total, we evaluated over 2600 prediction pipelines. We studied our final model by employing results from game theory in the form of Shapley values. Results indicate that predicting fluid intelligence from volume measurements is a challenging task with many challenges. We found that our final ensemble of 50 prediction pipelines associated larger parahippocampal gyrus volumes with lower fluid intelligence, and higher pons white matter volume with higher fluid intelligence.

1 Introduction

This paper describes our method submitted to the ABCD Neurocognitive Prediction Challenge 2019. The task of the challenge is to predict fluid intelligence solely from structural T1-weighted magnetic resonance images (MRI). The challenge uses data from the Adolescent Brain Cognitive Development (ABCD) Study.

In this approach, we first extract features from MRI scans and then use an automated machine learning approach for the prediction. For the feature extraction, we use volume measurements as provided by the challenge's organizers. For the prediction, we use an automated machine learning (AutoML) approach, as determining a good machine learning pipeline is a tedious and error-prone task for humans. A typical ML pipeline includes various types of preprocessing that can be applied to input features. Afterwards, an appropriate classifier needs to be selected and the optimal hyperparameters selected to achieve high predictive performance. The goal of AutoML is to automate the whole machine learning pipeline. A recent overview of AutoML approaches together with an analysis of the results of ChaLearn AutoML Challenges over the last four years

© Springer Nature Switzerland AG 2019
K. M. Pohl et al. (Eds.): ABCD-NP 2019, LNCS 11791, pp. 99–107, 2019.
https://doi.org/10.1007/978-3-030-31901-4_12

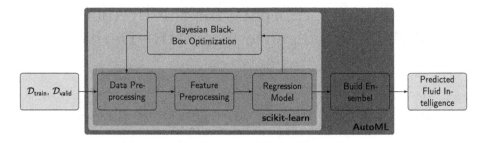

Fig. 1. Overview of our proposed AutoML pipeline for the prediction of fluid intelligence from T1-weighted MRI scans.

is given in [5]. AutoML has not yet been widely explored in the medical field, with PubMed listing only four articles [1,7,10,14]; none of which study MRI or neuroscience.

2   Data

Data was provided by The Adolescent Brain Cognitive Development(ABCD) Study [13], which recruited children aged 9–10. Participants were given access to T1-weighted MRI scans from 3,736 children for training, 415 children for validation, and 4,402 children for testing. Fluid intelligence scores were residualized to account for confounding due to sex at birth, ethnicity, highest parental education, parental income, parental marital status, and image acquisition site. Residualized fluid intelligence scores were provided for the training and validation data, but not for the test data. All data was obtained from the National Institute of Mental Health Data Archive.[1]

3   Methods

Our proposed pipeline for the prediction of fluid intelligence from T1-weighted MRI scans builds on the Automated Machine Learning (AutoML) framework summarized in Fig. 1. Scans were acquired according to the acquisition protocol of the Adolescent Brain Cognitive Development (ABCD) study protocol.[2] For parcellation of the brain and the estimation of volume of each region of interest, we relied on the work of the challenge's organizers.

3.1   Feature-Preprocessing

We used volume measurements of 122 regions of interest extracted by the challenge's organizers from each T1-weighted MRI scan based on the SRI24 atlas [15].[3] We normalized all volume measurements while accounting for out-

---

[1] https://nda.nih.gov/edit_collection.html?id=3104.

[2] https://abcdstudy.org/images/Protocol_Imaging_Sequences.pdf.

[3] See https://nda.nih.gov/data_structure.html?short_name=btsv01 for a full list of volumes.

liers by subtracting the median and dividing by the range between the 5% and 95% percentile. Thus, we reduce the impact of outliers and still obtain approximately centered features with equal scale. Finally, the provided residualized fluid intelligence scores in the training data where standardized to zero mean and unit variance; the same transformation as derived from the training data was applied to features and scores in the validation and test data. Additional pre-processing steps were selected without human interaction as described in the next section.

3.2   Automated Machine Learning

For the prediction of residualized fluid intelligence score, we used automated machine learning that leverages recent advances in Bayesian optimization, meta-learning, and ensemble construction. For every machine learning task, the fundamental problem is to decide which machine learning algorithm to use and whether and how to pre-process features. This task is extremely challenging, because there is no single algorithm that performs best on all datasets and the performance of machine learning methods depends to a large extent on their hyper-parameter settings, which can vary from one task to the next. Here, we use AutoML for the prediction of the residualized fluid intelligence score by producing test set predictions without human input within a given computational budget. Specifically, we employ Combined Algorithm Selection and Hyperparameter (CASH) optimization [3].

Let $\mathcal{A} = \{A^{(1)}, \ldots, A^{(R)}\}$ be a set of machine learning algorithms, and $\Lambda^{(j)}$ be the domain of the hyper-parameters of each algorithm. Further, we define $\mathcal{D}_{\text{train}} = \{(\mathbf{x}_1, y_1), \ldots, (\mathbf{x}_n, y_n)\}$ to be the training set, which we split into K cross-validation folds to obtain $\{\mathcal{D}_{\text{train}}^{(1)}, \ldots, \mathcal{D}_{\text{train}}^{(K)}\}$ and $\{\mathcal{D}_{\text{valid}}^{(1)}, \ldots, \mathcal{D}_{\text{valid}}^{(K)}\}$ with $\mathcal{D}_{\text{train}}^{(k)} = \mathcal{D}_{\text{train}} \backslash \mathcal{D}_{\text{valid}}^{(k)}$. For a particular hyper-parameter configuration Θ, we solve the CASH optimization problem

$$\underset{A^{(j)} \in \mathcal{A}, \Theta \in \Lambda^{(j)}}{\operatorname{argmin}} \quad \frac{1}{K} \sum_{k=1}^{K} \frac{1}{|\mathcal{D}_{\text{valid}}^{(k)}|} \sum_{i=1}^{|\mathcal{D}_{\text{valid}}^{(k)}|} \left(y_i - \hat{f}_{A_\Theta^{(j)}}(\mathbf{x}_i \mid \mathcal{D}_{\text{train}}^{(k)}), \right)^2 \qquad (1)$$

where $\hat{f}_{A_\Theta^{(j)}}(\mathbf{x}_i \mid \mathcal{D}_{\text{train}}^{(k)})$ denotes the prediction on the validation set of model $A^{(j)}$ with hyper-parameters Θ and trained on $\mathcal{D}_{\text{train}}^{(k)}$. This optimization problem can be solved via Sequential Model-based Algorithm Configuration (SMAC), a technique for Bayesian black-box optimization that uses a random-forest-based surrogate model [6]. The main idea of SMAC is to use the surrogate model to predict an algorithm's performance for a given hyper-parameter optimization. It is able to interpolate the performance of algorithms between observed hyper-parameter configurations and previously unseen configurations in the hyper-parameter space. Thus, it enables us to focus on promising hyper-parameter configurations.

We employed the auto-sklearn toolkit (version 0.5.0), which for a given user-provided computational budget in terms of run time and memory, auto-sklearn searches for the best machine learning pipeline to predict the residualized fluid intelligence score by combining components of the scikit-learn machine learning framework (version 0.18.2) [12]. Figure 1 depicts an overview of the AutoML framework. For data preprocessing, AutoML can choose from 11 algorithms for data transformations, such as principal component analysis. For feature preprocessing 6 feature-wise transformations are available, such as transforming each feature to have zero mean and unit variance. Finally, AutoML can choose from 13 regression models. After evaluating various machine learning pipelines, comprising data transformations, feature transformations, and regression model, the best M pipelines are combined via ensemble selection [2] to form the final prediction model. We used a budget that consisted of a total run time of 40 h, where each pipeline was limited to 6 min and 4 GB of memory. The final ensemble size was $M = 50$.

3.3   Feature Importance

While complex prediction pipelines are potentially powerful, their black-box nature is often a barrier for employing such a model in clinical research. We use Shapley values to explain the predictions of our final ensemble of prediction pipelines. Shapley values are a classic solution in game theory to determine the distribution of credits to players participating in a cooperative game [16,17]. They have first been proposed for linear models in the presence of multicollinearity [8]. A Shapley value assigns an importance value ϕ_j to each feature j that reflects its effect on the model's prediction. To compute this effect, retraining the model $f(\cdot)$ on all possible feature subsets $\mathcal{S} \subseteq \mathcal{F}\backslash\{j\}$ of all features $\mathcal{F}$ is necessary. Given a feature vector $\mathbf{x} \in \mathbb{R}^{|\mathcal{F}|}$, the j-th Shapley value can then be computed as the weighted average of all prediction differences:

$$\phi_j(\mathbf{x}) = \sum_{\mathcal{S} \subseteq \mathcal{F}\backslash\{j\}} \frac{|\mathcal{S}|!(|\mathcal{F}| - |\mathcal{S}| - 1)!}{|\mathcal{F}|!} \left(\hat{f}_{\mathcal{S} \cup \{j\}}(\mathbf{x}^{\mathcal{S} \cup \{j\}}) - \hat{f}_{\mathcal{S}}(\mathbf{x}^{\mathcal{S}}) \right), \qquad (2)$$

where $\hat{f}_{\mathcal{S}}(\mathbf{x}^{\mathcal{S}})$ denotes the prediction of a model trained and evaluated on the feature subset $\mathcal{S}$. The exact computation of Shapley values requires evaluating all $2^{|\mathcal{F}|}$ possible feature subsets, which is only reasonable when data consists of not more than a few dozen features. To address this problem, we employ the recently proposed SHAP (SHapley Additive exPlanations) values, which belong to the class of additive feature importance measures [9]. The exact computation of SHAP values is prohibitive, therefore we approximate SHAP values using the model-agnostic KernelSHAP approach proposed in [9]. To obtain a global measure of feature importance, we compute the average magnitude of SHAP values across all N subjects in the data:

$$\bar{\phi}_j = \frac{1}{N} \sum_{i=1}^{N} |\phi_j(\mathbf{x}_i)|. \qquad (3)$$

Table 1. Performance on training, validation and test set. MSE: mean squared error. MAE: mean absolute error.

	Subjects	MSE	MAE
Training	3,736	17.027	3.206
Validation	415	69.586	6.498
Test	4,402	94.010	—

Table 2. Summary of evaluated machine learning pipelines.

Description	Number
Algorithm runs	2608
Successful algorithm runs	2179
Crashed algorithm runs	6
Algorithms that exceeded the time limit	198
Algorithms that exceeded the memory limit	225

4   Results

The performance of the final ensemble is summarized in Table 1. It reveals that predicting residualized fluid intelligence from MRI-derived volume measurements is a challenging task. In particular, the proposed model struggles to reliably predict residualized fluid intelligence at the extremes of the distribution, i.e., very low or very high values. Consequently, we observe a relatively high mean squared error, which is an order of magnitude larger than the mean absolute error. Moreover, the large difference between the performance on the training data and the validation data indicates that overfitting seems to be an issue.

In total, we evaluated 2,608 machine learning algorithms (see Table 2). The components of our final ensemble of 50 machine learning pipelines is summarized in Table 3. Principal component analysis [11] was selected most often (15 times) for data pre-processing. The final ensemble was comprised of linear and non-linear regression models with ensembles of randomized regression trees [4] being selected most frequently (14 times). Looking at the top-performing pipelines in the ensemble, we noticed that combining principal component analysis with a tree-based ensemble was a frequently selected combination (5 out of the top 10 performing pipelines).

Next, we inspected which MRI-derived feature the model deems most important by computing SHAP values for each feature and subject in the training data. Figure 2 lists the top 20 features by mean absolute SHAP value ϕ. The top ranked feature is pons white matter volume ($\phi = 0.0183$), followed by left parahippocampal gyrus volume ($\phi = 0.0155$), and left lateral ventricle cerebral spinal fluid volume ($\phi = 0.0148$). However, we note that individual SHAP values are rather small, which is evidence that fluid intelligence is not strongly influenced by a single brain region, but a complex inter-relationship between

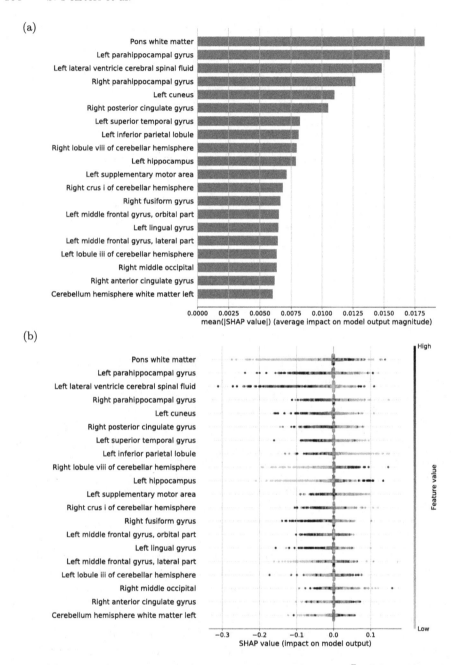

Fig. 2. (a) Top 20 features sorted by mean absolute SHAP value $\bar{\phi}_j$. (b) SHAP values of top 20 features for each subject in the training data. In each row SHAP values ϕ_j for each subject are plotted horizontally, stacking vertically to avoid overlap. Each dot is colored by the value of that feature, from low (blue) to high (red). If the impact of the feature on the model's prediction varies smoothly as its value changes then this coloring will also appear smooth. (Color figure online)

Table 3. Overview of selected components in the final ensemble of $M = 50$ pipelines selected by AutoML. Each pipeline consists of one data preprocessing step, one feature preprocessing step, and one regressor.

	Algorithm	Count
Data preprocessing	PCA	15
	Feature agglomeration	8
	Kernel PCA	8
	No preprocessing	6
	ICA	3
	Polynomial features	3
	Feature selection (Extra trees)	3
	Feature selection (percentile)	2
	Random trees embedding	1
	Nystroem sampler	1
Feature prepr.	Standardize	14
	None	13
	Normalize	7
	Min-max	6
	Quantile transformer	6
	Robust scaler	4
Regressor	Extra trees	14
	SGD	10
	Random forest	9
	Adaboost	5
	Decision tree	4
	Ridge regression	3
	Linear SVR	2
	ARD regression	1
	Gradient boosting	1
	k nearest neighbors	1

different regions. Individual, subject-specific SHAP values depicted in Fig. 2b indicate that larger left and right parahippocampal gyrus volume are associated with a decrease in fluid intelligence, while larger pons white matter volume is associated with an increase.

5   Conclusion

We proposed an AutoML model for the prediction of fluid intelligence from T1-weighted magnetic resonance images based on more than 2,600 evaluated machine learning pipelines. Our experiments demonstrate that it is challenging for our ensemble to reliably predict fluid intelligence from MRI scans. In particular, errors on the validation and test data were more than four times higher than on the training data, which is evidence for overfitting. We analyzed the final model's predictions using SHAP values. Results revealed that top ranked features still explain only a small fraction of the fluid intelligence score. Therefore, we concluded that current features derived from MRI are insufficient to robustly measure fluid intelligence. While current features are generic descriptors of the brain anatomy, we believe future research should focus on deriving tailor-made features from MRI, specific to the prediction of fluid intelligence, which could then be used to improve our understanding of the neurobiology underlying fluid intelligence.

Acknowledgements. This research was partially supported by the Bavarian State Ministry of Education, Science and the Arts in the framework of the Centre Digitisation.Bavaria (ZD.B).

References

1. Barreiro, E., Munteanu, C.R., Cruz-Monteagudo, M., Pazos, A., González-Díaz, H.: Net-net auto machine learning (AutoML) prediction of complex ecosystems. Sci. Rep. **8**(1), 12340 (2018)
2. Caruana, R., Niculescu-Mizil, A.: Ensemble selection from libraries of models. In: Proceedings of the 21st International Conference on Machine Learning, p. 18 (2004)
3. Feurer, M., Klein, A., Eggensperger, K., Springenberg, J., Blum, M., Hutter, F.: Efficient and robust automated machine learning. In: Advances in Neural Information Processing Systems 28, pp. 2962–2970 (2015)
4. Geurts, P., Ernst, D., Wehenkel, L.: Extremely randomized trees. Mach. Learn. **63**(1), 3–42 (2006)
5. Guyon, I., et al.: Analysis of the AutoML challenge series 2015–2018. In: Hutter, F., Kotthoff, L., Vanschoren, J. (eds.) Automated Machine Learning. TSSCML, pp. 177–219. Springer, Cham (2019). https://doi.org/10.1007/978-3-030-05318-5_10
6. Hutter, F., Hoos, H.H., Leyton-Brown, K.: Sequential model-based optimization for general algorithm configuration. In: Coello, C.A.C. (ed.) LION 2011. LNCS, vol. 6683, pp. 507–523. Springer, Heidelberg (2011). https://doi.org/10.1007/978-3-642-25566-3_40
7. Le, T.T., Fu, W., Moore, J.H.: Scaling tree-based automated machine learning to biomedical big data with a feature set selector. Bioinformatics, 1–7 (2019)
8. Lipovetsky, S., Conklin, M.: Analysis of regression in game theory approach. Appl. Stoch. Models Bus. Ind. **17**(4), 319–330 (2001)
9. Lundberg, S.M., Lee, S.I.: A unified approach to interpreting model predictions. In: Advances in Neural Information Processing Systems 30, pp. 4765–4774 (2017)
10. Orlenko, A., et al.: Considerations for automated machine learning in clinical metabolic profiling: altered homocysteine plasma concentration associated wtih metformin exposure. In: Pacific Symposium on Biocomputing, vol. 23. World Scientific (2017)
11. Pearson, K.: On lines and planes of closest fit to systems of points in space. Lond. Edinburgh Dublin Philos. Mag. J. Sci. **2**(11), 559–572 (1901)

12. Pedregosa, F., et al.: Scikit-learn: machine learning in python. J. Mach. Learn. Res. **12**, 2825–2830 (2011)
13. Pfefferbaum, A., et al.: Altered brain developmental trajectories in adolescents after initiating drinking. Am. J. Psychiatry **175**(4), 370–380 (2018)
14. Puri, M.: Automated machine learning diagnostic support system as a computational biomarker for detecting drug-induced liver injury patterns in whole slide liver pathology images. Assay Drug Dev. Technol. (2019)
15. Rohlfing, T., Zahr, N.M., Sullivan, E.V., Pfefferbaum, A.: The SRI24 multichannel atlas of normal adult human brain structure. Hum. Brain Mapp. **31**(5), 798–819 (2010)
16. Shapley, L.S.: A value for n-person games. Contrib. Theory Games **2**(28), 307–317 (1953)
17. Štrumbelj, E., Kononenko, I.: Explaining prediction models and individual predictions with feature contributions. Knowl. Inf. Syst. **41**(3), 647–665 (2014)

Predicting Fluid Intelligence from MRI Images with Encoder-Decoder Regularization

Lihao Liu, Lequan Yu$^{(\boxtimes)}$, Shujun Wang, and Pheng-Ann Heng

Department of Computer Science and Engineering,
The Chinese University of Hong Kong, Shatin, Hong Kong
lqyu@cse.cuhk.edu.hk

Abstract. In this paper, we develop a 3D convolutional neural network to predict the fluid intelligence from T1-weighted MRI images by adding an encoder-decoder regularization. Considering that cerebellar volume is often highly correlated to intelligence of an individual, we propose to incorporate this morphological information into the framework for fluid intelligence prediction by utilizing an encoder-decoder regularization for brain structure segmentation simultaneously. Specifically, we first train an encoder-decoder network to generate the brain segmentation mask, where the discriminative morphological feature of the brain volume can be learned. Then, we reuse the encoder path of the network as the prediction network backbone for final fluid intelligence prediction by adding an additional regression part to predict the fluid intelligence value. The proposed framework is able to learn the discriminative relationship between the morphological information of brain structures and the intelligence score for more accurate prediction.

1 Introduction

Determining the neural mechanisms underlying general intelligence is fundamental to understanding cognitive development, how this relates to real-world health outcomes, and how interventions (education, environment) might improve outcomes through adolescence and into adulthood[1]. Among different types of general intelligence, fluid intelligence is a major factor in measuring general intelligence [3], which can be measured via the NIH Toolbox Neurocognition battery [1] and from which demographic confounding factors (*e.g.*, sex, and age) are removed. It is an emerging topic to use machine learning based methods to predict fluid intelligence from medical images via data-driven manner. However, direct prediction of fluid intelligence from the brain MRI Images is often challenging due to the lack of determinant factor. Furthermore, direct regression from the brain volumes is easy to overfit the training data with lower performance on testing samples.

---

[1] https://sibis.sri.com/abcd-np-challenge/.

© Springer Nature Switzerland AG 2019
K. M. Pohl et al. (Eds.): ABCD-NP 2019, LNCS 11791, pp. 108–113, 2019.
https://doi.org/10.1007/978-3-030-31901-4_13

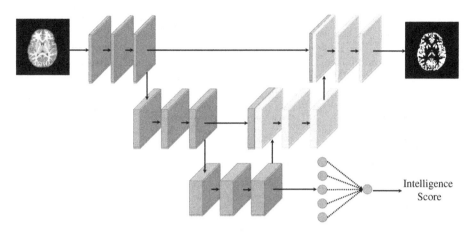

Fig. 1. The schematic illustration of the overall framework. The blue blocks and the white blocks denote the 3D convolutional layers in the encoder and decoder part, respectively. We then add an additional regression layer (blue dot) for the regression task. (Color figure online)

In this paper, we develop a 3D convolutional neural network (CNN) based framework to predict the fluid intelligence from T1-weighted MRI images. The 3D CNN is able to fully incorporate the 3D information and geometric cues of the MRI images for effective fluid intelligence prediction. Although lack of determinate factor, intelligence is found to be significantly correlated with intracranial, cerebral, temporal lobe, hippocampal, and cerebellar volume [2]. Therefore, to improve the prediction accuracy, we propose to incorporate the morphological information into the framework for fluid intelligence prediction. In particular, we utilize an encoder-decoder regularization to facilitate the model to learn a more discriminative morphological feature by conducting the brain structure simultaneously. We propose a two-stage training scheme to train the whole framework. We first train an encoder-decoder-like network to conduct the brain structure segmentation task, and then we reuse the encoder part as the prediction network backbone to conduct the fluid intelligence prediction. In the first stage, we train the model using the MR brain volumes and its corresponding structure masks from the training and validation subset. By conducting the segmentation task, the network can learn a generalized feature for the fluid intelligence prediction. Next, we discard the decoder part and fine-tune the encoder part with an additional regression branch to predict the fluid intelligence value, in which the MR brain volumes and the fluid intelligence scores are used. The encoder part with the regression branch (blue part in Fig. 1) is used as our final 3D CNN architecture for fluid intelligence prediction. This two-stage training pipeline alleviates the overfitting problem of the network when directly regressing the fluid intelligence from MR images.

Table 1. The network architecture of our proposed method. To reduce the size of feature maps, we set the stride of the first convolutional layer at each stage in the encoder path as 2. "[ ]" indicates a basic residual block (not bottleneck) in which the *conv* denotes a combination of a concolutional layer, a batch normalization layer and a relu activation layer. While fc denotes the fully-connected layer.

	Encoder		Decoder	
	Operations	Output size	Operations	Output size
	$conv\ 7 \times 7 \times 7, 16$	$48 \times 48 \times 32 \times 16$	$deconv\ 3 \times 3 \times 3, 3$ $conv\ 3 \times 3 \times 3, 1$	$96 \times 96 \times 64 \times 1$
Stage1	$\begin{bmatrix} conv\ 3 \times 3 \times 3, 16 \\ conv\ 3 \times 3 \times 3, 16 \end{bmatrix} \times 2$	$24 \times 24 \times 16 \times 16$	$deconv\ 3 \times 3 \times 3, 16$ $concatenation$ $\begin{bmatrix} conv\ 3 \times 3 \times 3, 16 \\ conv\ 3 \times 3 \times 3, 16 \end{bmatrix} \times 2$	$24 \times 24 \times 16 \times 16$
Stage2	$\begin{bmatrix} conv\ 3 \times 3 \times 3, 32 \\ conv\ 3 \times 3 \times 3, 32 \end{bmatrix} \times 2$	$12 \times 12 \times 6 \times 32$	$deconv\ 3 \times 3 \times 3, 32$ $concatenation$ $\begin{bmatrix} conv\ 3 \times 3 \times 3, 32 \\ conv\ 3 \times 3 \times 3, 32 \end{bmatrix} \times 2$	$12 \times 12 \times 6 \times 32$
Stage3	$\begin{bmatrix} conv\ 3 \times 3 \times 3, 64 \\ conv\ 3 \times 3 \times 3, 64 \end{bmatrix} \times 2$	$6 \times 6 \times 3 \times 64$	—	—
Regression module	$flatten$ $fc\ 6912 \times 256$ $fc\ 256 \times 1$	1	—	—

2 Methodology

2.1 Network Architecture

Our proposed framework is based on 3D convolutional neural network to fully incorporate the 3D information of the MRI volumes. To improve the generality capability of network and learn more discriminative semantic features, we further utilize an encoder-decoder regularization scheme to train our model in a two-stage manner.

Figure 1 demonstrate the overall framework of our method. We first train an encoder-decoder-like architecture, which takes an MR brain volume as the input and outputs the segmentation result in an end-to-end manner. We use multiple convolutional layers to generate a set of 3D convolutional feature maps with multiple resolutions; see the blue blocks in the left part of Fig. 1. Then, the deepest highly semantic feature maps with the lowest resolution (the bottom row in Fig. 1) are repeatedly enlarged by the deconvolutional layers (decoder part) and concatenated with the corresponding feature maps from the encoder part via the skip connection. Next, we reconstruct the segmentation mask of the input volume, and update the weights in the encoder-decoder by calculating the cross-entropy loss between the predicted mask and the ground truth mask. The details of architecture is shown in Table 1.

In the second training stage, we discard the decoder part and fine-tune the learned weights in the encoder part. We further add a regression module behind the encoder, which contains one fully connected layer without any activation layer, to predict the intelligence score. We update the regression module by

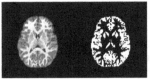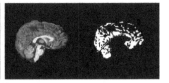

Fig. 2. Different views of the 3D Data samples (MR brain volumes and its segmentation masks) from training dataset.

calculating the mean square error loss between the ground truth intelligence score and the predicted score.

2.2 Training Details

To accelerate the training process, we initialize the parameters of all the convolutional layers in our network with the "uniform" initialization method. We adopt the Adam optimizer [4] to optimize the network with a weight decay of 0.0001 and a batch size of eight for both the first and second training stages. We set the learning rate as 0.0001, and periodically reduce it by multiplying 0.9 in every 1,000 iterations, and the training process is terminated after 10,000 iterations without early stop for both stages. Our method is implemented with Tensorflow and DLTK toolbox [5].

3   Experiments

3.1   Dataset

In the ABCD challenge dataset [6], the training, validation and testing subset contain 3739, 415 and 4402 individual subjects, respectively. Each subject contains a 3D MR brain volume and a corresponding structure segmentation mask, as shown in Fig. 2. These images are with the uniform volume size ($240 \times 240 \times 240$). Besides, the training and validation subset also have a *pre-residual intelligence score* for each individual subjects.

3.2   Data Pre-processing and Experimental Setup

To facilitate the training procedure, we conducted some pre-processing steps for each 3D brain volume. We first resized the brain volume to $120 \times 120 \times 120$ using bilinear interpolation. Then we center cropped a $120 \times 120 \times 90$ region from the resized volume, considering the z dimension contains less information than the x and y dimensions. We also performed "whitening" operation to normalize the intensity to zero mean and unit variance. To increase the total amount of training data and enhance the robustness of the network, we used random flipping and random cropping as data augmentation in the training process. Specifically, we randomly cropped a $96 \times 96 \times 64$ region out $120 \times 120 \times 90$ original brain volume as the input of the network during the training.

During the two-stage training process, we first use the brain volumes and its corresponding segmentation masks from training and validation subsets to train the encoder-decoder architecture without updating the regression part. In this step, we merged the labeled brain structures and regarded the segmentation as a binary segmentation task. In the second stage, we fixed the weights of the encoder part and update the regression part using the brain volume and the provided intelligence score. In the testing phase, we take the MR brain volume with the same pre-processing steps as input and directly output the regressed pre-residual intelligence score.

3.3   Evaluation Metrics and Results

Encoder-Decoder Segmentation Results. To validate whether the encoder-decoder learned the morphological features, we use dice coefficient score as the evaluation metric. The dice coefficient score computes the region based similarity between the predicted segmentation result and the ground truth segmentation mask:

$$Dice\,(P,G) = \frac{2 \times |P \cap G|}{|P| + |G|} ,\qquad(1)$$

where P denotes the predicted segmentation result, G denotes the ground truth segmentation mask, $|P \cap G|$ denotes the overlapped region between P and G, and $|P| + |G|$ represents the union region. Noted, a larger $Dice$ indicates a better segmentation result. Our trained encoder-decoder achieved a $Dice$ of 0.9767 in the validation dataset. While in the testing dataset, we also achieved a similar $Dice$ of 0.9465, which indicates the learned convolutional layers can extract discriminative features from MRI volumes.

Fluid Intelligence Prediction Results. In the testing/validation phase, we used the ten-crops to obtain the final results. Specifically, we randomly cropped ten regions ($96 \times 96 \times 64$) from the pre-processed images ($120 \times 120 \times 90$) and separately made a prediction for each region with the trained network. Then, we averaged the ten predicted scores as the final output score for one input image. With only the training dataset, our method achieved an MSE error of 71.5679 at the validation dataset. In the final testing phase, we merged the training volumes and validation volumes to the whole framework, and our method achieved an MSE error of 102.2498 on the testing data.

4   Conclusion

This paper presents a 3D convolutional neural network for fluid intelligence prediction from T1-weighted MRI images. We employ an encoder-decoder segmentation regularization to learn discriminative morphological feature of the brain volume for better fluid intelligence value prediction. The proposed two-stage framework can reduce the overfitting of the network when directly regressing fluid intelligence values. The proposed framework can be generalized to other related regression problems.

Acknowledgment. The work described in this paper was supported by a grant from the Research Grants Council of Hong Kong Special Administrative Region, China (Project No. CUHK 14225616).

References

1. Akshoomoff, N., et al.: VIII. NIH toolbox cognition battery (CB): composite scores of crystallized, fluid, and overall cognition. Monogr. Soc. Res. Child Dev. **78**(4), 119–132 (2013)
2. Andreasen, N.C., et al.: Intelligence and brain structure in normal individuals. Am. J. Psychiatry **150**, 130 (1993)
3. Carroll, J.B.: Human Cognitive Abilities: A Survey of Factor-Analytic Studies. Cambridge University Press, Cambridge (1993)
4. Kingma, D.P., Ba, J.: Adam: a method for stochastic optimization. arXiv preprint arXiv:1412.6980 (2014)
5. Pawlowski, N., et al.: DLTK: state of the art reference implementations for deep learning on medical images. arXiv preprint arXiv:1711.06853 (2017)
6. Pfefferbaum, A., et al.: Altered brain developmental trajectories in adolescents after initiating drinking. Am. J. Psychiatry **175**(4), 370–380 (2017)

ABCD Neurocognitive Prediction Challenge 2019: Predicting Individual Residual Fluid Intelligence Scores from Cortical Grey Matter Morphology

Neil P. Oxtoby[1]([✉])[iD], Fabio S. Ferreira[1,2][iD], Agoston Mihalik[1,2][iD],
Tong Wu[1,2][iD], Mikael Brudfors[1,3][iD], Hongxiang Lin[1][iD], Anita Rau[1,3][iD],
Stefano B. Blumberg[1][iD], Maria Robu[1,4][iD], Cemre Zor[1,2][iD], Maira Tariq[1][iD],
Mar Estarellas Garcia[1][iD], Baris Kanber[5][iD], Daniil I. Nikitichev[1,3][iD],
and Janaina Mourão-Miranda[1,2][iD]

[1] Centre for Medical Image Computing (CMIC), Department of Computer Science
and Department of Medical Physics and Biomedical Engineering,
University College London, Gower Street, London WC1E 6BT, UK
n.oxtoby@ucl.ac.uk
[2] Max Planck UCL Centre for Computational Psychiatry and Ageing Research,
University College London, Gower Street, London WC1E 6BT, UK
[3] The Wellcome Centre for Human Neuroimaging,
University College London, Gower Street, London WC1E 6BT, UK
[4] Wellcome/EPSRC Centre for Interventional and Surgical Sciences (WEISS),
University College London, Gower Street, London WC1E 6BT, UK
[5] Department of Clinical and Experimental Epilepsy, Queen Square Institute
of Neurology, University College London, Gower Street, London WC1E 6BT, UK

Abstract. We predicted fluid intelligence from T1-weighted MRI data
available as part of the ABCD NP Challenge 2019, using morphological similarity of grey-matter regions across the cortex. Individual structural covariance networks (SCN) were abstracted into graph-theory metrics averaged over nodes across the brain and in data-driven commu-
nities/modules. Metrics included degree, path length, clustering coefficient, centrality, rich club coefficient, and small-worldness. These features derived from the training set were used to build various regression
models for predicting residual fluid intelligence scores, with performance
evaluated both using cross-validation within the training set and using
the held-out validation set. Our predictions on the test set were generated with a support vector regression model trained on the training
set. We found minimal improvement over predicting a zero residual fluid
intelligence score across the sample population, implying that structural
covariance networks calculated from T1-weighted MR imaging data provide little information about residual fluid intelligence.

Keywords: Support vector regression · Fluid intelligence · MRI ·
Structural covariance networks · Graph theory features

N. P. Oxtoby and F. S. Ferreira—These authors contributed equally to this work.

1   Introduction

Establishing the neurobiological mechanisms underlying intelligence is a key area of research in neuroscience [1]. A strong correlation has been observed between cognitive ability measured at a very young age with socioeconomic status [2], as well as longevity and health [3] at an older age. Moreover, intelligence has been shown to be very stable from young to old age in the same individuals [4,5]. Thus understanding the mechanisms of cognitive abilities has implications for health of the general population and can be used to enhance such abilities, for example through education or environment [6].

Neuroimaging plays a key role in advancing our knowledge of the neural mechanisms of intelligence. Several brain-imaging studies have shown the link between brain features and intelligence, including a positive correlation with cortical volume and thickness, specifically in the frontal and temporal regions [7–11]. A link has also been observed between intelligence and the structural integrity of white matter [12] and the function integrity of the temporal, frontal and parietal cortices [13]. Studies have also involved both adult and children [14,15]. The ABCD NP Challenge asks the question "How predictable is fluid intelligence from brain imaging data?" To answer this, we took a data-driven, exploratory approach of trying many models and image-based features—starting with a hackathon led by the UCL Centre for Medical Image Computing (CMIC). CMIC aims to make an impact on key medical challenges facing 21st-century society through performing world-leading research on problems in medical imaging and image-analysis. Our expertise extends from feature extraction/generation through to image-based modelling [16,17], machine learning [18,19], and beyond. The hackathon took place one afternoon in February 2019 and involved researchers across research groups in UCL CMIC, in addition to colleagues from the affiliated UCL Wellcome Centre for Human Neuroimaging, UCL Department of Clinical and Experimental Epilepsy, and Max Planck UCL Centre for Computational Psychiatry and Ageing Research. Regular followup progress meetings followed the hackathon.

The brain is a complex organ widely touted as operating as a cliquish small-world network [20], although this may not be the whole story [21]. The ABCD NP Challenge lacks the diffusion MRI data necessary to estimate anatomical connectivity via tractography. However, it is possible to quantify morphological similarity of an individual's cortex using a graph called a "structural covariance network" (SCN), which can be used to distinguish between clinical groups [22]. We calculate SCNs for each individual in the ABCD NP Challenge data set and input them as features to train predictive models of residual fluid intelligence (rFIQ).

The paper is structured as follows. The next section describes the challenge data and our methods. Section 3 presents our results which we discuss in Sect. 4 then conclude.

2   Methods

2.1   Data

The ABCD NP Challenge data consists of a cross-section of imaging data and intelligence scores for children aged 9–10 years. The T1-weighted MRI data was acquired using the protocol detailed on the challenge website [23] and in [24], and split into training ($N = 3739$), validation ($N = 415$), and test ($N = 4515$) sets. The training and validation sets also include scores of fluid intelligence, which the ABCD Study measures using the NIH Toolbox Neurocognition battery [25]. For the challenge, fluid intelligence was residualized to remove dependence upon brain volume, data collection site, age at baseline, sex at birth, race/ethnicity, highest parental education, parental income, and parental marital status. While we understand the motivation—the challenge is to predict intelligence from imaging—this pre-residualization choice in the challenge design is somewhat limiting because it completely removes any ability to include covariance of these factors with image-based features. The MRI data provided was already in pre-processed form. Pre-processing included skull-stripping, removing noise, correcting for field inhomogeneities [26,27] and affine alignment of all images to the SRI24 adult brain atlas [28]. The SRI24 segmentations and corresponding volumes were also provided. Unsurprisingly, the regional volumes were not predictive of a target that had been adjusted for total brain volume.

2.2   Structural Covariance Network Features

It has been shown that cortical morphology is predictive of cognitive deficits in individuals with Alzheimer's disease [22]. We wanted to explore whether the same could be said for predicting intelligence, so we generated a structural covariance network (SCN) following [29] (code available on GitHub) for each individual in the ABCD NP Challenge data set. The SCN is a graph where the nodes are small cortical regions (3 voxels cubed) and the edges quantify structural similarity (morphology) between nodes. From each SCN we generated nodal graph-theory features using the Brain Connectivity Toolbox [30], which were then averaged across the brain and also within each of the largest three modules (communities) of the graph. We also considered measures of variation in these features (standard deviation and median-absolute deviation). Our 26 features include small-worldness, rich club coefficient, path length, node degree, clustering coefficient, and betweenness centrality (Table 1). See Fig. 1 for a graphical representation of the pipeline.

Generating approximately ten thousand SCNs and corresponding graph-theory features is an intensive computational task. When the pipeline failed for a given individual, or time was not permitting (such as the late addition of 868 additional test subjects), this resulted in missing data. For these few individuals ($\leq 8\%$: Table 1) we inserted a prediction of zero (nominally the mean).

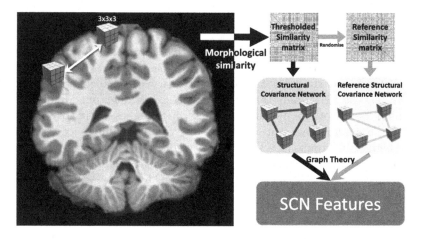

Fig. 1. Structural covariance network feature-generation pipeline.

2.3   Predictive Models

We trained two models to predict rFIQ from features based on morphological similarity. The first was the event-based model (EBM) of progression [17,31]. The second was support-vector regression (SVR) [32]. We trained each model on data from the training set, and assessed performance using MSE on the validation set (Table 1). The best-performing model (SVR) was used to generate our submission to the challenge: predictions for the test set.

The EBM learns a discrete sequence of progression events from normal/low state to abnormal/high. It was designed for neurodegenerative diseases but can be applied to any monotonic phenomenon. Here we define low rFIQ as more than one standard deviation (std) below the mean and high rFIQ as more than one std above the mean. If rFIQ is a monotonic function of structural covariance, then the EBM should be able to find a probabilistic sequence of events that represent this function. "Events" are structural covariance graph-theory features, and they must differ statistically between low-rFIQ and high-rFIQ for them to be included in the model—otherwise they contain no "signal" for this trajectory. We excluded features that "did not pass" ($p > 0.10$) the Mann-Whitney U test of the null hypothesis that the distributions (low/high rFIQ) are equal. EBM stage and rFIQ score was input into a Kernel Ridge Regression model (default parameters, scikit-learn: [33]) to make the predictions.

The SVR was run in PRoNTo version 3 (Pattern Recognition for Neuroimaging Toolbox; http://www.mlnl.cs.ucl.ac.uk/pronto) [34]—a software toolbox of pattern recognition techniques for the analysis of neuroimaging data. Model performance on the training set was assessed using 5-fold nested cross-validation (i.e. the internal and external loops had 5 folds) to optimise the penalty parameter C (we use 6 different logarithmically-spaced values: 0.01, 0.1, 1, 10, 100 and 1000) and compute the MSE per fold, which were averaged across folds to compute the final prediction error (Table 2).

3 Results

We included 26 SCN graph-theory features that represent morphological similarity across the cortex. Table 1 summarises the features we derived from the T1 images, and the level of completeness in each challenge data set (see Sect. 2.3). For the EBM, only three features passed through our Mann-Whitney U test filter (see Methods): small-worldness, betweenness centrality (median), and degree. Even for these features, there was very little difference between the low- and high-rFIQ groups (see Table 1), with Cohen's d effect sizes of $-0.11/0.06$ (small-world), $0.07/-0.10$ (degree), and $-0.09/0.009$ (centrality) in the training/validation sets. In light of the opposing effect direction (signs), the model's poor generalisation performance is unsurprising (see Table 2).

Table 1. Descriptive values for 26 SCN graph-theory features across training, validation, and test sets. Values are: mean (std). Missing data was due to feature generation failure (see Methods): training set 96% complete; validation 94%; test 92%. Notes: Centrality = Betweenness centrality; Clustering = Clustering coefficient.

Whole network features	Training (N = 3579 of 3739)	Validation (N = 390 of 415)	Test (N = 4156 of 4515)
Small-world	1.68 (0.03)	1.68 (0.02)	1.68 (0.02)
Rich club – median	0.29 (0.01)	0.29 (0.01)	0.29 (0.01)
– mad	0.11 (0.03)	0.11 (0.01)	0.11 (0.01)
Path length – median	2.48 (0.03)	2.48 (0.01)	2.48 (0.01)
– std	1.15 (0.03)	1.15 (0.02)	1.16 (0.02)
Degree – median	1050 (45)	1052 (45)	1053 (40)
– mad	295 (19)	294 (15)	294 (15)
Centrality – median	6584 (171)	6590 (150)	6578 (152)
– mad	5153 (157)	5157 (118)	5158 (117)
Clustering – median	0.53 (0.01)	0.53 (0.01)	0.53 (0.01)
– mad	0.063 (0.005)	0.063 (0.005)	0.063 (0.005)
Community 1/2/3 features			
Avg. degree – 1	995 (233)	1004 (232)	995 (239)
– 2	996 (240)	997 (235)	999 (239)
– 3	1019 (242)	1004 (243)	1014 (238)
Avg. degree z-score (all)	0.2 (0.1)	0.2 (0.1)	0.2 (0.1)
Avg. path length (all)	1.5 (0.1)	1.5 (0.1)	1.5 (0.1)
Centrality – 1	6020 (3750)	6180 (3800)	6100 (3780)
– 2	6290 (3790)	6030 (3740)	6290 (3770)
– 3	6660 (3830)	6550 (3760)	6520 (3810)
Clustering – 1	0.53 (0.06)	0.53 (0.06)	0.52 (0.06)
– 2	0.53 (0.06)	0.53 (0.06)	0.53 (0.06)
– 3	0.53 (0.06)	0.53 (0.06)	0.52 (0.06)

For the SVR model, two features were most important: small-worldness (weight $w = 11.42$); and clustering coefficient in community 2 ($w = 6.04$). Among the next most important were average path length and other clustering coefficients.

Table 2 shows our prediction results for both models: mean-squared errors and Pearson's squared correlation coefficient for training and validation. It is clear that both the approaches did not generalise well under validation. Our submission to the challenge (SVR) was positioned near the middle of the testing leaderboard with MSE = 93.8335 (Table 2).

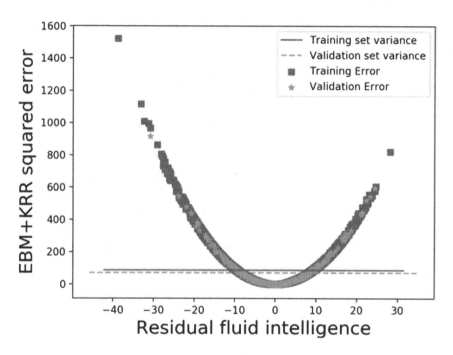

Fig. 2. Training and validation errors for the EBM approach.

4   Discussion

The ABCD NP Challenge was certainly challenging. Our MSE for predicting residual fluid intelligence was only nominally better than simply predicting zero, i.e., the mean. This implies that the residual fluid intelligence is not explainable by graph theory features derived from structural covariance networks. We found similar results for all combinations of models and features attempted during and after our hackathon—from basic regression to deep learning. Moreover, the validation leader board (see challenge website) demonstrated that other entries into the challenge had similarly meagre performance improvement on simply predicting the mean.

Table 2. Mean-squared error (MSE) and correlation for the predictive models. For reference, the variance of the training set was 85.85 and the validation set was 71.53.

Prediction method	Training set		Validation set		Test set
	MSE	Correlation	MSE	Correlation	MSE
SVR	85.82	0.02	71.19	0.01	**93.8335**
EBM+KRR	85.46	0.001	71.58	0.003	N/A

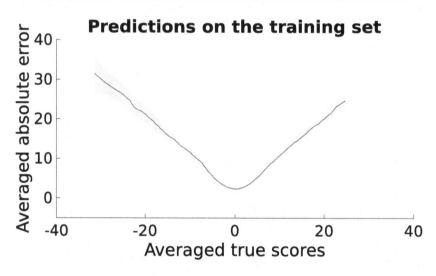

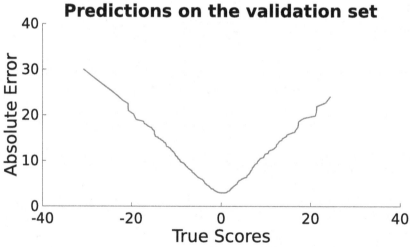

Fig. 3. SVR prediction errors: (left) training set using 5-fold CV; (right) validation set.

While the residualization process precluded the use of models that include covariance of the residualization factors [18] with image-based features, it is difficult to say whether or not this would have improved the results dramatically. Including variables in the residualization procedure that are correlated with the predicted variable is likely to remove important variability in the data leading to predictive models with low performance [35].

5 Conclusion

Based on our results, and those on the validation leaderboard for the challenge, we are inclined to conclude that structural imaging is probably incapable of predicting more than a couple of points worth of residual fluid intelligence.

Acknowledgements. This study was funded by the UCL Centre for Medical Image Computing and UK EPSRC platform grant "Medical image computing for next-generation healthcare technology" (EP/M020533/1) and supported by researchers at the National Institute for Health Research University College London Hospitals Biomedical Research Centre. FSF is funded by a PhD scholarship awarded by Fundacao para a Ciencia e a Tecnologia (SFRH/BD/120640/2016). NPO acknowledges support from the NIHR UCLH Biomedical Research Centre and the EuroPOND project— This project has received funding from the European Union's Horizon 2020 research and innovation programme under grant agreement No. 666992. AM, CZ, TW, and JM-M acknowledge funding from the Wellcome Trust under grant number WT102845/Z/13/Z.

References

1. Goriounova, N.A., Mansvelder, H.D.: Genes, cells and brain areas of intelligence. Front. Hum. Neurosci. **13**, 44 (2019). https://doi.org/10.3389/fnhum.2019.00044
2. Foverskov, E., Mortensen, E.L., Holm, A., Pedersen, J.L.M., Osler, M., Lund, R.: Socioeconomic position across the life course and cognitive ability later in life: the importance of considering early cognitive ability. J. Aging Health **31**(6), 947–966 (2017). https://doi.org/10.1177/0898264317742810
3. Lam, N.H., et al.: Effects of Altered Excitation-Inhibition Balance on Decision Making in a Cortical Circuit Model. bioRxiv 100347 (2017). https://doi.org/10.1101/100347
4. Deary, I.J., Strand, S., Smith, P., Fernandes, C.: Intelligence and educational achievement. Intelligence **35**(1), 13–21 (2007). https://doi.org/10.1016/j.intell.2006.02.001
5. Deary, I.J., Pattie, A., Starr, J.M.: The stability of intelligence from age 11 to age 90 years: the lothian birth cohort of 1921. Psychol. Sci. **24**(12), 2361–2368 (2013). https://doi.org/10.1177/0956797613486487
6. Gottfredson, L.S.: Why g matters: the complexity of everyday life. Intelligence **24**(1), 79–132 (1997). https://doi.org/10.1016/S0160-2896(97)90014-3
7. Hulshoff Pol, H.E., et al.: Genetic contributions to human brain morphology and intelligence. J. Neurosci. **26**(40), 10235–10242 (2006). https://doi.org/10.1523/JNEUROSCI.1312-06.2006

8. Narr, K.L., et al.: Relationships between IQ and regional cortical gray matter thickness in healthy adults. Cereb. Cortex **17**(9), 2163–2171 (2007). https://doi.org/10.1093/cercor/bhl125

9. Choi, Y.Y., et al.: Multiple bases of human intelligence revealed by cortical thickness and neural activation. J. Neurosci. **28**(41), 10323–10329 (2008). https://doi.org/10.1523/JNEUROSCI.3259-08.2008

10. Karama, S., et al.: Cortical thickness correlates of specific cognitive performance accounted for by the general factor of intelligence in healthy children aged 6 to 18. NeuroImage **55**(4), 1443–1453 (2011). https://doi.org/10.1016/j.neuroimage.2011.01.016

11. Jung, R.E., Haier, R.J.: The Parieto-Frontal Integration Theory (P-FIT) of intelligence: converging neuroimaging evidence. Behav. Brain Sci. **30**(2), 135–154 (2007). https://doi.org/10.1017/S0140525X07001185

12. Penke, L., et al.: Brain white matter tract integrity as a neural foundation for general intelligence. Mol. Psychiatry **17**, 1026 (2012). https://doi.org/10.1038/mp.2012.66

13. Wang, L., et al.: Alterations in cortical thickness and white matter integrity in mild cognitive impairment measured by whole-brain cortical thickness mapping and diffusion tensor imaging. Am. J. Neuroradiol. **30**(5), 893–899 (2009). https://doi.org/10.3174/ajnr.A1484

14. Muetzel, R.L., et al.: White matter integrity and cognitive performance in school-age children: a population-based neuroimaging study. NeuroImage **119**, 119–128 (2015). https://doi.org/10.1016/J.NEUROIMAGE.2015.06.014

15. Yu, C., et al.: White matter tract integrity and intelligence in patients with mental retardation and healthy adults. NeuroImage **40**(4), 1533–1541 (2008). https://doi.org/10.1016/j.neuroimage.2008.01.063

16. Oxtoby, N.P., Alexander, D.C.: for the EuroPOND consortium: Imaging plus X: multimodal models of neurodegenerative disease. Curr. Opin. Neurol. **30**(4), 371–379 (2017). https://doi.org/10.1097/WCO.0000000000000460

17. Young, A.L., et al.: A data-driven model of biomarker changes in sporadic Alzheimer's disease. Brain **137**(9), 2564–2577 (2014). https://doi.org/10.1093/brain/awu176

18. Schrouff, J., Monteiro, J.M., Portugal, L., Rosa, M.J., Phillips, C., Mourão-Miranda, J.: Embedding anatomical or functional knowledge in whole-brain multiple kernel learning models. Neuroinformatics **16**(1), 117–143 (2018). https://doi.org/10.1007/s12021-017-9347-8

19. Blumberg, S.B., Tanno, R., Kokkinos, I., Alexander, D.C.: Deeper image quality transfer: training low-memory neural networks for 3D images. In: Frangi, A.F., Schnabel, J.A., Davatzikos, C., Alberola-López, C., Fichtinger, G. (eds.) MICCAI 2018. LNCS, vol. 11070, pp. 118–125. Springer, Cham (2018). https://doi.org/10.1007/978-3-030-00928-1_14

20. Bassett, D.S., Bullmore, E.: Small-world brain networks. Neuroscientist **12**(6), 512–523 (2006). https://doi.org/10.1177/1073858406293182

21. Bassett, D.S., Sporns, O.: Network neuroscience. Nat. Neurosci. **20**(3), 353–364 (2017). https://doi.org/10.1038/nn.4502

22. Tijms, B.M., et al.: Single-subject grey matter graphs in Alzheimer's Disease. PLoS ONE **8**(3), e58921 (2013). https://doi.org/10.1371/journal.pone.0058921

23. https://abcdstudy.org/images/Protocol_Imaging_Sequences.pdf

24. Casey, B.J., et al.: The adolescent brain cognitive development (ABCD) study: imaging acquisition across 21 sites. Dev. Cogn. Neurosci. **32**, 43–54 (2018). https://doi.org/10.1016/j.dcn.2018.03.001

25. Akshoomoff, N., et al.: VIII. NIH toolbox cognition battery (CB): composite scores of crystallized, fluid, and overall cognition. Monogr. Soc. Res. Child Dev. **78**(4), 119–132 (2013). https://doi.org/10.1111/mono.12038
26. Hagler, D.J., et al.: Image processing and analysis methods for the Adolescent Brain Cognitive Development Study. bioRxiv 457739 (2018). https://doi.org/10.1101/457739
27. Pfefferbaum, A., et al.: Altered brain developmental trajectories in adolescents after initiating drinking. Am. J. Psychiatry **175**(4), 370–380 (2018). https://doi.org/10.1176/appi.ajp.2017.17040469
28. Rohlfing, T., Zahr, N.M., Sullivan, E.V., Pfefferbaum, A.: The SRI24 multichannel atlas of normal adult human brain structure. Hum. Brain Mapp. **31**(5), 798–819 (2010). https://doi.org/10.1002/hbm.20906
29. Lawrie, S.M., Tijms, B.M., Willshaw, D.J., Seriès, P.: Similarity-based extraction of individual networks from gray matter MRI scans. Cereb. Cortex **22**(7), 1530–1541 (2012). https://doi.org/10.1093/cercor/bhr221
30. Rubinov, M., Sporns, O.: Complex network measures of brain connectivity: uses and interpretations. NeuroImage **52**(3), 1059–1069 (2010). https://doi.org/10.1016/j.neuroimage.2009.10.003
31. Fonteijn, H.M., et al.: An event-based model for disease progression and its application in familial Alzheimer's disease and Huntington's disease. NeuroImage **60**(3), 1880–1889 (2012). https://doi.org/10.1016/j.neuroimage.2012.01.062
32. Drucker, H., Burges, C.J.C., Kaufman, L., Smola, A., Vapnik, V.: Support vector regression machines. In: Proceedings of the 9th International Conference on Neural Information Processing Systems, NIPS 1996, pp. 155–161. MIT Press, Cambridge (1996). http://dl.acm.org/citation.cfm?id=2998981.2999003
33. Pedregosa, F., et al.: Scikit-learn: machine learning in python. J. Mach. Learn. Res. **12**, 2825–2830 (2011)
34. Schrouff, J., et al.: PRoNTo: pattern recognition for neuroimaging toolbox. Neuroinformatics **11**(3), 319–37 (2013). https://doi.org/10.1007/s12021-013-9178-1
35. Rao, A., Monteiro, J.M., Mourao-Miranda, J.: Alzheimer's disease initiative: predictive modelling using neuroimaging data in the presence of confounds. NeuroImage **150**, 23–49 (2017). https://doi.org/10.1016/j.neuroimage.2017.01.066

Ensemble Modeling of Neurocognitive Performance Using MRI-Derived Brain Structure Volumes

Leo Brueggeman⬤, Tanner Koomar⬤, Yongchao Huang, Brady Hoskins,
Tien Tong, James Kent, Ethan Bahl, Charles E. Johnson, Alexander Powers,
Douglas Langbehn, Jatin Vaidya, Hans Johnson, and Jacob J. Michaelson(✉)

University of Iowa, Iowa City, IA 52240, USA
jacob-michaelson@uiowa.edu

Abstract. Prediction of cognitive performance from brain structural imaging data is a challenging machine learning topic. Participating in the ABCD Neurocognitive prediction challenge (2019), we implemented several machine learning models to solve this problem. Our results show superior performance from models relying on boosted decision trees and we find benefit from using two different sets of derived brain volumetric features. Lastly, across all models, we report an increase in performance by ensembling several different model types together in a final layer.

Keyword: ABCD-NP challenge machine learning intelligence

1 Introduction

Prediction of neurocognitive performance from brain imaging data is a long-standing problem at the intersection of neuroscience and computer vision. Early research in this field robustly established that overall brain volume correlates strongly with intelligence (r = 0.24) [12]. However, conclusions regarding how specific brain region volumes might impact intelligence have been sparse.

Several existing methods which predict fluid intelligence or other cognitive metrics using structural brain features that extend beyond total brain volume have been published. For instance, gray matter density in several cortical regions including the orbitofrontal cortex and cingulate gyrus has been correlated with the general intelligence quotient (IQ) in adolescents [2]. While cortical structures have been a focal point in intelligence research, there is emerging evidence that other brain structures (such as the hippocampus) may play a role in intelligence [14]. More recent machine-learning approaches have used up to several hundred brain region volumes to predict intelligence. One such approach was able to predict intelligence with an average correlation coefficient of 0.718 using support vector regression [15].

L. Brueggeman and T. Koomar—Denotes equal authorship.

© Springer Nature Switzerland AG 2019
K. M. Pohl et al. (Eds.): ABCD-NP 2019, LNCS 11791, pp. 124–132, 2019.
https://doi.org/10.1007/978-3-030-31901-4_15

While many of the existing studies have shown great performance in their ability to predict intelligence, they have several key drawbacks. First, sample sizes in MRI-based studies have been limited, typically involving only a few hundred individuals at most. Second, most predictive studies have not accounted for the effects of age, demographics, or total brain volumes in their ability to predict intelligence using specific brain region volumes. This step is critical, as many of these variables are correlated with IQ [2], and may also be correlated with specific brain region volumes. Without accounting for these broad differences, the contribution of specific brain regions to overall intelligence cannot be meaningfully interpreted. The ABCD Neurocognitive prediction challenge addresses both of these limitations by introducing the largest brain MRI-intelligence cohort to date, with appropriate corrections for known confounders.

Using this data, our team trained numerous machine learning models to predict the confounder-residualized intelligence scores. While models such as support vector machines have been used before with this problem [15], we undertook a broad survey of available machine learning models in order to find those best suited to the task. After systematically ranking method performance, we evaluated the information gain through ensembling multiple machine learning models in a final training step, balancing the strengths and weaknesses of each approach. Using this strategy, we highlight machine learning models particularly well suited to the high-dimensional problem of intelligence prediction and provide evidence for the benefits of ensemble learning.

2   Methods

2.1   Source of Data

Data for the ABCD challenge was accessed through the NDAR data portal, under the ABCD-Neurocognitive Prediction Challenge 2019 datasets. Specifically, we accessed DICOM images from brain MRI scans, MRI-derived brain volumetric measurements provided as part of the challenge [11], and neurocognitive performance metrics. In total, MRI and neurocognitive performance metrics were provided for 3739 and 415 individuals in the training and validation datasets. The test set consisted of MRI data and brain volume metrics for 4515 individuals.

2.2   Data Processing

Brain MRI DICOM images were processed using the open source medical image analysis suite, BRAINSTools (https://github.com/BRAINSia/BRAINSTools). This pipeline produced an additional 210 volumetric measurements of brain anatomical structures which were used for modeling, in addition to the volumes provided by ABCD. In brief, the BRAINSAutoWorkup uses the best-rated T1 images for spatial normalization based on landmarks such as the anterior/posterior commissures and the mid-sagittal plane. The remaining scans are then rigidly aligned to the best-rated T1 image, and processed with the automated bias-field correction (BRAINSABC) algorithm. Scans are then segmented for subcortical structures using an automated segmentation framework. The BRAINSTools derived brain volumes were combined with the provided brain volumes (mapped to the SRI atlas) and used for further modeling.

2.3 SRI and BRAINSTools Feature Performance Comparison

To directly compare the usefulness of SRI and BRAINSTools volumes, we fit a Random Forest model using either SRI-volumes, BRAINSTools-volumes, or SRI- and BRAINSTools-volumes as features. The model was fit using the caret package [8] in R [13] in repeated cross-validation (k = 10, n = 2; same splits) with a random-search strategy for hyperparameter optimization (length = 20). Cross-validated performance of the best performing models in the training set were used for evaluation. Feature importance within the combined SRI- and BRAINSTools-volumes Random Forest model was computed via permutation importance.

2.4 Model Training Protocol

Using the training dataset, we split off 739 individuals to form a validation set. This was done via random split and using a Kolmogorov-Smirnov test to minimize difference between the intelligence and volumetric measurements between the validation set and the full training dataset. We chose to form our own validation dataset after finding a substantial left-shift in the intelligence scores from the validation dataset provided as part of the challenge (training mean = 0.0555, validation mean = −0.503). Machine learning models were then fit to the training dataset using the caret package [8] in R [13]. Hyperparameter optimization was performed using a random search strategy in repeated cross-validation. The parameters for each model are depicted in (Table 1).

Trained models were filtered for better than random predictive performance on the validation set (only successful models shown), using the correlation coefficient as the performance metric. An elastic-net ensemble model was then fit on the validation dataset using the predictions of the six successful models as features. A random search strategy was used in 10-fold cross validation, repeated 10 times, to find the optimal hyperparameters for the elastic-net model. This ensemble model was then used to predict on the test dataset, used for performance evaluation by the ABCD-Neurocognitive predictive challenge.

3 Results

3.1 Brain Region Segmentation Quality

Our approach to the ABCD-NP challenge started with an evaluation of all provided data. The accuracy of SRI-derived brain segmentation volumes [11] was of primary importance, as this was the main feature-space for our machine learning models. When visualizing brain MRIs with overlaid brain region segmentations

Table 1. Hyperparameters chosen based on 5-fold repeated cross-validation.

Model	Hyperparemeters
xgbtree	rounds $= 700$; max_depth $= 16$; $\eta = 0.005$; $\gamma = 5$; columns_sampled $= 0.2$; min_child_weight $= 1$; subsample $= 0.66$
MARS	nprune $= 11$; degree $= 1$
xgbtree_impute	rounds $= 71$; max_depth $= 6$; $\eta = 0.0142$; $\gamma = 7.3$; columns_sampled $= 0.685$; min_child_weight $= 0$; subsample $= 0.709$
enet	$\alpha = 0.2628$; $\lambda = 2.608$
bstTree	mstop $= 5$; maxdepth $= 1$; $\nu = 0.1$
treebag	none

we observed that several sampled scans had underestimated subcortical structure boundaries. Specifically, we observed large regions of the putamen, thalamus, and caudate which fell outside of the provided region boundaries (Fig. 1A). Due to this discrepancy, we reprocessed the full dataset of brain MRI DICOM images using a recent segmentation pipeline, BRAINSTools [1,3–7]. This yielded an additional 210 brain volume measurements for each subject, in addition to the 122 SRI-based brain volumes. BRAINSTools volumes included white matter, gray matter, subcortical, cerebellar, and CSF segmentations, some of which were overlapping other segmentations and/or measured in both hemispheres. Visualizing the subcortical segmentations from BRAINSTools showed improved segmentation for the putamen and thalamus, with minor improvements in the caudate (Fig. 1B).

3.2   Brain Region Segmentation Strategy Comparison

With two brain region volume datasets available for modeling, we next tested whether one was a superior feature-space when modeling neurocognitive performance. Random Forest models were fit in cross-validation on the training dataset using either SRI, BRAINSTools, or SRI and BRAINSTools features. Performance was evaluated by Pearson's correlation coefficient (see Sect. 2.3 for further training details). The best performing model in this analysis relied on both the BRAINSTools and SRI features (Fig. 2A), suggesting added value from our additional data processing step.

Looking at the feature importance within this model, 10 of the 11 top features in this model were from the BRAINSTools dataset (Fig. 2B). By far the most important feature was the BRAINSTools variable corresponding to unsegmented white matter, which corresponds to white matter voxels not fitting within any of the particular white matter segmentations. Other important features included multiple white matter segmentations, several subcortical structures such as the hippocampus and amygdala, and several cortical structures such as the superior-frontal and middle temporal cortex. Despite having a greater number of features

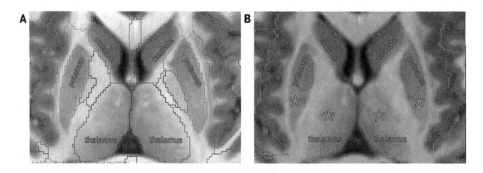

Fig. 1. Subcortical region segmentations by BRAINStools (A) and the provided segmentations mapped to the SRI atlas (B). Asterisks in (B) indicate portions of the labeled regions not properly detected.

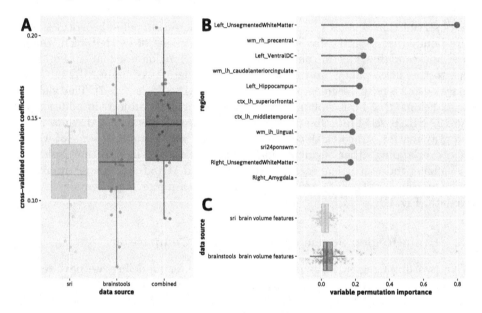

Fig. 2. Cross-validated performance in the training set from a random forest using the two brain region segmentation strategies to predict neurocognitive performance (A). Top feature importance values (permutation-based) from the model using both segmentation strategy volumes (B). Teal points are features from BRAINSTools, and light purple points are features from the SRI segmentation. Distribution of all feature importance values from both segmentation strategies (C).

(BRAINSTools = 210, SRI = 122), BRAINSTools features showed a significantly higher variable permutation importance (one-sided Mann-Whitney U test P-value = 0.001196; Fig. 2C).

3.3   Model Performance in Training and Validation Datasets

Using both the SRI and BRAINSTools datasets as features, we next trained a variety of machine learning models to predict neurocognitive performance (see training approach in Sect. 2.4). Tree-based models survived an initial screen of models capable of predicting intelligence with better than random performance, with the addition of an elastic net (enet) and a bagged multivariate adaptive regression spline model (MARS). Applying these models to the validation dataset, we found a high correlation between predicted values between the different models (Fig. 3A). This high correlation was especially present between models from the same class, such as the two extreme gradient boosted models (xgbTree) which differed in their data preprocessing. While the correlation between predictions from most models was high (0.4+), the correlation between predictions and the neurocognitive scores in the validation set was much lower (0.05–0.11). Similarly, mean cross-validated correlation with neurocognitive scores in the training set also ranged from 0.05–0.13 (Fig. 3B). Comparing the training set and validation set performance, we found that the elastic net model had the biggest discrepancy in scores, indicating less generalizability to unseen data. Conversely, we found that the single best performing model in the validation dataset was a bagged classification and regression tree (treebag) model, with a Pearson's correlation coefficient of 0.105 (P-value = 0.0042). Unlike many other models assessed, the treebag model had very similar training and validation correlation coefficients, indicating the data had not been overfit.

Our final submission to the ABCD neurocognitive prediction challenge was an elastic net ensemble model using all models shown in Fig. 3B as features. This model was fit on the validation dataset, and applied to the provided test set for final evaluation by the challenge organizers. Specifically, an elastic net model was fit in 10-fold repeated (n = 10) cross validation using a random search hyperparameter optimization strategy (fraction = 0.22, lambda = 7.32). The final ensemble elastic net model showed similar performance (Pearson's r = 0.110; P-value = 0.0027) in the validation dataset (estimated through cross validation) to the treebag individual model. Notably, the final ensemble model assigned a non-zero coefficient to all features, except for the elastic net model. The single most important feature was the treebag model predictions, followed closely by the MARS model.

4   Discussion

Predicting intelligence metrics from brain imaging and structural data remains a challenging problem. By correcting for many factors known to influence intelligence metrics, results from the ABCD neurocognitive prediction challenge are more focused in their predictive scope than many prior studies of intelligence. Because of this, the correlation coefficients we report from our training and validation datasets are significantly below what was previously published [15], when these factors were not taken into consideration. Despite accounting for these

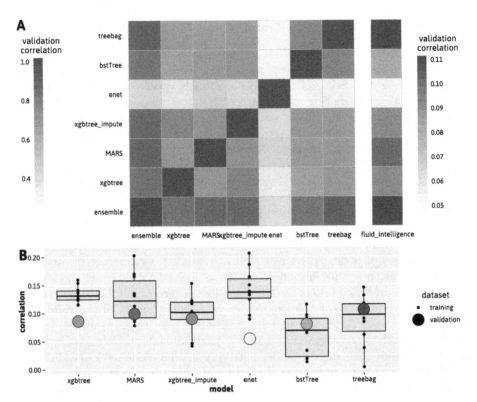

Fig. 3. Pearson's correlation coefficient of predictions in the validation dataset between all models (A-left). Correlation between predictions and true neurocognitive performance values in the validation dataset (A-right). Cross-validated correlation coefficients in the training dataset (black points) from each model, visualized with performance in the validation dataset (blue points; matches data from A-right). (Color figure online)

major factors known to influence intelligence, we still report appreciable signal in the ability of brain region volumes to predict neurocognitive performance.

Central to our models performance was an additional round of processing and segmentation of the ABCD MRI data. A model using both segmentation strategies in isolation or combination showed that the best performance was achieved when all features were made available. This result suggests that both the SRI-based and BRAINSTools segmentation strategies added unique signal which increased performance, despite expanding the total number of features used.

By evaluating the importance of brain region features in predicting neurocognitive performance, we found several white matter related volumes which were highly informative. This represents a shift from previous reports, where the emphasis was largely placed on grey matter segmentation volumes [15]. Overall, white matter volume has previously been reported to significantly predict intelligence, but with less predictive power than total brain volume or grey matter volume [10]. This difference may be because total brain volume, the majority of which is grey matter [9], was regressed out of the neurocognitive performance scores predicted here.

In our final trained model, we saw benefit from using an ensemble based approach. Importantly, the ensemble model achieved the greatest Pearson correlation coefficient through cross-validation on the validation dataset, outperforming all individual models. With five of the six supplied models receiving a non-negative coefficient, this suggests that individual models each provided some unique signal which was useful in predicting neurocognitive performance. The ensemble model achieved a cross-validated MSE on the ensemble model's training set (i.e. our internal validation dataset) of 80.45, and a MSE of 71.14 on the ABCD-NP provided validation dataset. On the test set, the ensemble model achieved a MSE of 92.49 (second place).

In sum, we report significant prediction of neurocognitive performance metrics in the ABCD dataset through use of an elastic net ensemble model with a cross-validated Pearson's correlation coefficient of 0.110. To improve upon these results, we suspect that the use of datasets with both structural and functional MRI data will allow even greater sensitivity of predictions. Importantly, as learned through this challenge, accounting for the influence of factors such as total brain volume and demographic information makes the prediction of neurocognitive performance more challenging, but also allows for more robust conclusions about the effect of brain region volume on intelligence.

References

1. Forbes, J.L., Kim, R.E.Y., Paulsen, J.S., Johnson, H.J.: An open-source label atlas correction tool and preliminary results on huntingtons disease whole-brain MRI atlases. Front. Neuroinformatics 10 (2016). https://doi.org/10.3389/fninf.2016.00029

2. Frangou, S., Chitins, X., Williams, S.C.: Mapping IQ and gray matter density in healthy young people. NeuroImage 23(3), 800–805 (2004). https://doi.org/10.1016/j.neuroimage.2004.05.027

3. Ghayoor, A., Paulsen, J.S., Kim, R.E.Y., Johnson, H.J.: Tissue classification of large-scale multi-site MR data using fuzzy k-nearest neighbor method. In: Styner, M.A., Angelini, E.D. (eds.) Medical Imaging 2016: Image Processing. SPIE, March 2016. https://doi.org/10.1117/12.2216625

4. Ghayoor, A., Vaidya, J.G., Johnson, H.J.: Robust automated constellation-based landmark detection in human brain imaging. NeuroImage 170, 471–481 (2018). https://doi.org/10.1016/j.neuroimage.2017.04.012

5. Kim, E.Y., Johnson, H.J.: Robust multi-site MR data processing: iterative optimization of bias correction, tissue classification, and registration. Front. Neuroinformatics 7, (2013). https://doi.org/10.3389/fninf.2013.00029

6. Kim, R.E.Y., Lourens, S., Long, J.D., Paulsen, J.S., Johnson, H.J.: Preliminary analysis using multi-atlas labeling algorithms for tracing longitudinal change. Front. Neurosci. 9 (2015). https://doi.org/10.3389/fnins.2015.00242

7. Kim, R.E.Y., Nopoulos, P., Paulsen, J., Johnson, H.: Efficient and extensible work-flow: reliable whole brain segmentation for large-scale, multi-center longitudinal human MRI analysis using high performance/throughput computing resources. In: Oyarzun Laura, C., et al. (eds.) CLIP 2015. LNCS, vol. 9401, pp. 54–61. Springer, Cham (2016). https://doi.org/10.1007/978-3-319-31808-0_7
8. Kuhn, M.: Building predictive models in r using the caret package. J. Stat. Softw. Articles **28**(5), 1–26 (2008). https://doi.org/10.18637/jss.v028.i05
9. Lüders, E., Steinmetz, H., Jüncke, L.: Brain size and grey matter volume in the healthy human brain. NeuroReport **13**(17), 2371–2374 (2002). https://doi.org/10.1097/00001756-200212030-00040
10. Luders, E., Narr, K.L., Thompson, P.M., Toga, A.W.: Neuroanatomical correlates of intelligence. Intelligence **37**(2), 156–163 (2009). https://doi.org/10.1016/j.intell.2008.07.002
11. Pfefferbaum, A., et al.: Altered brain developmental trajectories in adolescents after initiating drinking. Am. J. Psychiatry **175**(4), 370–380 (2018). https://doi.org/10.1176/appi.ajp.2017.17040469
12. Pietschnig, J., Penke, L., Wicherts, J.M., Zeiler, M., Voracek, M.: Meta-analysis of associations between human brain volume and intelligence differences: how strong are they and what do they mean? SSRN Electron. J. (2014). https://doi.org/10.2139/ssrn.2512128
13. R Core Team: R: A Language and Environment for Statistical Computing. R Foundation for Statistical Computing, Vienna, Austria (2019). https://www.R-project.org/
14. Supekar, K., et al.: Neural predictors of individual differences in response to math tutoring in primary-grade school children. Proc. Natl. Acad. Sci. **110**(20), 8230–8235 (2013). https://doi.org/10.1073/pnas.1222154110
15. Wang, L., Wee, C.Y., Suk, H.I., Tang, X., Shen, D.: MRI-based intelligence quotient (IQ) estimation with sparse learning. PLoS ONE **10**(3), e0117295 (2015). https://doi.org/10.1371/journal.pone.0117295

ABCD Neurocognitive Prediction Challenge 2019: Predicting Individual Fluid Intelligence Scores from Structural MRI Using Probabilistic Segmentation and Kernel Ridge Regression

Agoston Mihalik[1,2], Mikael Brudfors[1,3], Maria Robu[1,4],
Fabio S. Ferreira[1,2], Hongxiang Lin[1], Anita Rau[1,4], Tong Wu[1,2],
Stefano B. Blumberg[1], Baris Kanber[5], Maira Tariq[1],
Mar Estarellas Garcia[1], Cemre Zor[1,2], Daniil I. Nikitichev[1,4],
Janaina Mourão-Miranda[1,2], and Neil P. Oxtoby[1(✉)]

[1] Centre for Medical Image Computing (CMIC), Department of Computer Science
and Department of Medical Physics and Biomedical Engineering,
University College London, Gower Street, London WC1E 6BT, UK
n.oxtoby@ucl.ac.uk
[2] Max Planck UCL Centre for Computational Psychiatry and Ageing Research,
University College London, Gower Street, London WC1E 6BT, UK
[3] The Wellcome Centre for Human Neuroimaging,
University College London, Gower Street, London WC1E 6BT, UK
[4] Wellcome/EPSRC Centre for Interventional and Surgical Sciences (WEISS),
University College London, Gower Street, London WC1E 6BT, UK
[5] Department of Clinical and Experimental Epilepsy,
Queen Square Institute of Neurology, University College London,
Gower Street, London WC1E 6BT, UK

Abstract. We applied several regression and deep learning methods to
predict fluid intelligence scores from T1-weighted MRI scans as part of
the ABCD Neurocognitive Prediction Challenge 2019. We used voxel
intensities and probabilistic tissue-type labels derived from these as features to train the models. The best predictive performance (lowest mean-
squared error) came from kernel ridge regression ($\lambda = 10$), which produced a mean-squared error of 69.7204 on the validation set and 92.1298
on the test set. This placed our group in the fifth position on the validation leader board and first place on the final (test) leader board.

Keywords: Kernel ridge regression · Fluid intelligence · PRoNTo ·
MRI · Convolutional neural networks · Hackathon

A. Mihalik, M. Brudfors, J. Mourão-Miranda and N. P. Oxtoby–These authors contributed equally to this work.

K. M. Pohl et al. (Eds.): ABCD-NP 2019, LNCS 11791, pp. 133–142, 2019.
https://doi.org/10.1007/978-3-030-31901-4_16

1   Introduction

Establishing the neurobiological mechanisms underlying intelligence is a key area of research in neuroscience [1, 2]. General intelligence at a young age is predictive of later educational achievement, occupational attainment, and job performance [3–6]. Moreover, intelligence in childhood or early adulthood is associated with health outcomes later in life as well as mortality [5, 7–10]. Thus, understanding the mechanisms of cognitive abilities in children potentially has important implications for society and can be used to enhance such abilities, for example through targeted interventions such as education and the management of environmental risk factors [4, 11].

Neuroimaging can play a key role in advancing our understanding of the neural mechanisms of cognitive ability. Several brain-imaging studies have shown that total brain volume is the strongest brain imaging derived predictor of general intelligence [12–14] ($r \approx 0.3 - 0.4$). To a somewhat lesser degree, regional cortical volume and thickness differences in the frontal, temporal, and parietal lobes have also been linked to intelligence [12, 13, 15–17]. Converging neuroimaging evidence led to the proposal of the parieto-frontal integration theory [18] whereby a distributed network of brain regions is responsible for the individual variability in cognitive abilities. This theory is also supported by human lesion studies [19, 20].

The ABCD-NP Challenge 2019 asked the question *"Can we predict fluid intelligence from T1-weighted MRI?"* We took an exploratory, data-driven approach to answering this question—a hackathon organised by our local research centres: the UCL Centre for Medical Image Computing (CMIC) and Wellcome/EPSRC Centre for Interventional and Surgical Sciences (WEISS). Our centres aim to address key medical challenges facing 21st-century society through world-leading research in medical imaging, medical image analysis, and computer-assisted interventions. Our expertise extends from feature extraction/generation through to image-based modelling [21], machine learning [22–24], and beyond. The hackathon involved researchers across research groups in our centres, in addition to colleagues from the affiliated Wellcome Centre for Human Neuroimaging and the Department of Clinical and Experimental Epilepsy at UCL. The hackathon took place on an afternoon in February 2019, after which we followed up with regular progress meetings.

In this paper we report our findings for predicting fluid intelligence in 9/10-year-olds from T1-weighted MRI using machine learning regression and deep learning methods (convolutional neural networks—CNNs). Our paper is structured as follows. The next section describes the challenge data and our methods. Section 3 presents our results, which we discuss in Sect. 4 before concluding.

2   Methods

2.1   Data

The ABCD-NP Challenge data consists of pre-processed T1-weighted MRI scans and fluid intelligence scores for children aged 9–10 years. The imaging protocol can be found in [25]. Pre-processing included skull-stripping, noise removal, correction for field inhomogeneities [26,27], and affine alignment to the SRI24 adult brain atlas [28]. SRI24 segmentations and corresponding volumes were also provided.

The cohort was split into training ($N = 3739$), validation ($N = 415$), and test ($N = 4515$) sets. The training and validation sets also include scores of fluid intelligence, which are measured in the ABCD Study using the NIH Toolbox Neurocognition battery [29]. For the challenge, fluid intelligence was residualised to remove linear dependence upon brain volume, data collection site, age at baseline, sex at birth, race/ethnicity, highest parental education, parental income and parental marital status.

2.2   Features Derived from the Data

We trained the models to predict fluid intelligence both from the provided T1-weighted images (voxel intensity) as well as voxel-wise feature maps generated from these images using a probabilistic segmentation approach. There are many different methods to extract various features from T1-weighted images [30], such as tissue-type labels obtained from probabilistic segmentations. These segmentations can be constructed in a way to capture not only the relative tissue composition in a voxel, but also information about shape differences between individuals. This requires mapping each subject to a common template—a fundamental technique of computational anatomy [31]. Here we constructed such a template from all available T1-weighted MRI scans ($S = train + validation + test = 3,739 + 415 + 4,515 = 8,669$), which generated normalised (non-linearly aligned to a common mean) tissue segmentations for each subject.

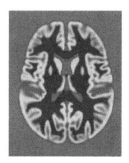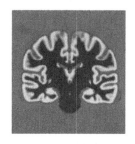

Fig. 1. Template generated from fitting the generative model to all of the subjects in the ABCD population ($S = 8,669$). Green corresponds to grey matter tissue, blue to white matter, and red to other. (Color figure online)

We used a generative model [32] to probabilistically segment each T1-weighted MRI in the challenge data set into three tissue types: grey-matter, white-matter, other—see Fig. 1. The in-house model[1] used here contains key improvements over the one in [32]: (1) we place a smoothing prior on the template; (2) we obtain better initial values by first working on histogram representations of the images; (3) we normalise over population image intensities in a principled way, within the model; (4) we place a prior on the proportions of each tissue, which is also learned during training. There are two types of normalised segmentations: non-modulated and modulated. Modulated segmentations include the relative shape change when aligning to the common template.

Seven features per voxel were considered: T1-weighted intensity plus our six derived features corresponding to modulated and non-modulated probabilities for each tissue type. All images, including the feature maps from probabilistic segmentations, were spatially smoothed with a Gaussian kernel of 12 mm FWHM [30] and masked to remove voxels outside of the brain.

2.3   Predicting Fluid Intelligence: Machine Learning Regression

We explored several machine learning regression algorithms of varying complexity, including Multi-Kernel Learning (MKL) [33], Kernel Ridge Regression (KRR) [34], Gaussian Process Regression (GPR) [35] and Relevance Vector Machines [36]. The inputs to these models consisted of different concatenated combinations of our seven voxel-wise features described in Sect. 2.2. Analyses were run in *PRoNTo version 3* [22] (http://www.mlnl.cs.ucl.ac.uk/pronto), a software toolbox of pattern recognition techniques for the analysis of neuroimaging data, as well as custom-written code.

In our preliminary analyses, we trained different combinations of regression algorithms and features using 5-fold cross-validation within the training set to select the best combination of algorithm and features as measured by lowest cross-validated mean-squared error (MSE). We then retrained the best-performing model using the entire training set and used this trained model to generate predictions of fluid intelligence scores for the validation and test sets. Our best-performing model was KRR using all six voxel-wise derived features (tissue-type probabilities) concatenated into an input feature vector of length ~ 1.7 million per individual. We set the regularization hyperparameter to $\lambda = 10$ [34], which was optimised through 5-fold nested cross-validation within the training set in preliminary analyses.

We investigated robustness/stability of our KRR model using modified jack-knife resampling (80/20 train/test split). Explicitly, we trained the model on a random subsample of 80% of the training set and generated predictions for both the held out 20% of the training set and the full validation set. We repeated this procedure 1000 times to generate confidence bounds on performance (MSE).

2.4   Predicting Fluid Intelligence: Convolutional Neural Networks

Separately, we explored the use of CNNs. The motivation was to incorporate spatial information that is not explicitly modelled as features. The CNNs were

---

[1] Available from https://github.com/WCHN/segmentation-model.

trained directly on the pre-processed T1-weighted images. Similarly to previous work that predicted brain age, Alzheimer's disease progression, or brain regions from MRI scans [37–39] we applied various layers of 3D-convolutional kernels with filter size of $3 \times 3 \times 3$ voxels on down-sampled images with dimensions of $61 \times 61 \times 61$ voxels. We trained and validated multiple neural networks including those in [24]. Our best performing CNN (lowest MSE) consisted of four convolutional layers and three fully-connected layers followed by dropout layers with a probability of 0.5. The first six layers were activated by rectified linear units. The convolutional layers were followed by batch normalization and max-pooling operations. We used the Adam optimizer with an initial learning rate of 10^{-5} and a decay of 10^{-5}, and we stopped training at the 100th epoch. To evaluate the network, we randomly sampled 10 subsets of 1870 subjects from the training set, applied them to train 10 CNN models and evaluated performance on the validation set. We report MSE averaged over the ten passes.

3   Results

Our best-performing regression model was KRR using our six derived voxel-wise features as input. This produced MSE = 69.72 on the validation set. By comparison, our best-performing CNN achieved MSE = 70.82 (average MSE = 73.40) on the validation set. However, we observed that the better-performing CNNs on the training set did not generalise well to the validation set, possibly due to over-fitting, or mismatch between training and validation data (as suggested by the considerable difference in variances).

Table 1 shows MSE and Pearson's correlation coefficient (mean $\pm$ std) for training and validation predictions from our top-performing methods. We found that KRR performed the best and we used this model to generate our challenge submission, which resulted in MSE = 92.13 on the test set.

Table 1. Model performance for the training and validation sets.

Method	Training set		Validation set		Test set
	MSE	Correlation	MSE	Correlation	MSE
KRR	77.64	0.1427	**69.72**	0.0311	**92.13**
CNN	69.10 ± 3.13	0.1913 ± 0.0423	73.40 ± 2.13	-0.0202 ± 0.0296	N/A
	(best: 64.39)	(best: 0.2542)	(best: 70.82)	(best: 0.0157)	

Figure 2 shows model robustness results from modified jackknife resampling (1000 repetitions, 80/20 split). Confidence in our predicted MSE on the validation set is within ± 1 residual IQ point.

Figure 3 shows the residual fluid intelligence score prediction errors for KRR on the training and validation sets. The V shape of the curves shows that smaller residual scores have lower errors, since the model is predicting close to the mean value.

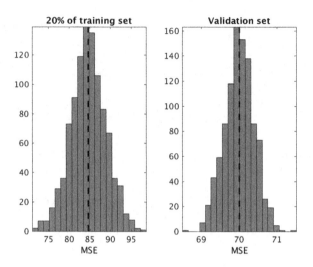

Fig. 2. Model robustness for KRR using modified jackknife re-sampling. MSE distributions for 1000 predictions: 20% of the training set (left); full validation set (right); for models trained on the other 80% of the training set. Red line indicates the mean. (Color figure online)

4   Discussion

We found that predicting residual fluid intelligence from structural MRI images is challenging. The correlation between predicted and actual intelligence scores was low for all methods we tested ($r \approx 0.05 - 0.15$). This contrasts with previous studies for predicting (non-residualised) fluid intelligence, which have demonstrated that both total brain volume and regional cortical volume/thickness differences are relatively strong predictors ($r \approx 0.3 - 0.4$) [12–14].

The lower predictive performance we observed might be influenced by the residualisation, which prevents modelling of covariance between the residualisation factors and the image-based features. Moreover, there is evidence that including variables in the residualisation procedure that are correlated with the regression targets/labels is likely to remove important variability in the data leading to predictive models with low performance [40].

We note that previous studies used small sample sizes (of order 10–100), whilst the ABCD-NP Challenge dataset comprised a very large-scale dataset (of order 1000–10000). Whereas subject recruitment for small samples tend to be well controlled, resulting in homogeneous sample characteristics, large samples are more heterogeneous by nature, thus predictive models are more challenging to build. Accordingly, a recent study has demonstrated that the accuracy of classification results tends to be smaller for larger sample sizes [41].

Our image-based features were voxel-wise probabilistic tissue-type labels: grey matter, white matter, and other. Beyond tissue-type labels, there might be value in investigating other features generated by the generative segmentation

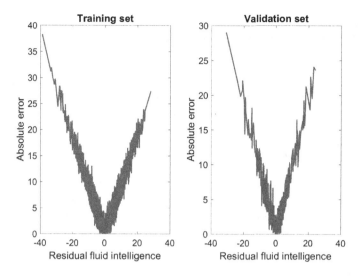

Fig. 3. Absolute prediction errors against residual fluid intelligence scores for KRR (training set on the left, validation set on the right).

model. One such interesting feature is scalar momentum [30], which has been shown to be predictive for a range of different problems [30].

Finally, we mention a possible limitation of our approach. In recent years, it has become standard practice to create study-specific group templates in neuroimaging [31], especially in Voxel-Based Morphometry (VBM) analyses. Originally, this approach was proposed for group analysis using mass univariate statistics (e.g., statistical parametric mapping). However, one needs to exercise caution when applying such an approach in machine learning, as it might lead to slightly optimistic predictions by creating dependence across the overall dataset. In order to avoid this potential issue, one would need to create templates based only on the training set. This might be computationally challenging when cross-validation strategies are used. To the best of our knowledge, no studies have investigated whether study-specific templates indeed result in slightly inflated predictions, and it remains an interesting question for future work.

5   Conclusion

Our paper presents the winning method for the ABCD Neurocognitive Prediction Challenge 2019. We found that kernel ridge regression outperformed more complex models, such as convolutional neural networks, when predicting residual fluid intelligence scores for the challenge dataset using our custom tissue-type features derived from the preprocessed T1-weighted MRI. The correlation between the predicted and actual scores is very low ($r = 0.03$ for the KRR on the validation set), implying that the association between structural images and residualised fluid intelligence scores is low. It may be that structural images contain very little information on residualised fluid intelligence, but further study is warranted.

Acknowledgements. This study was funded by the UCL Centre for Medical Image Computing and UK EPSRC platform grant "Medical image computing for next-generation healthcare technology" (EP/M020533/1) and supported by researchers at the National Institute for Health Research University College London Hospitals Biomedical Research Centre. FSF is funded by a PhD scholarship awarded by Fundacao para a Ciencia e a Tecnologia (SFRH/BD/120640/2016). NPO acknowledges support from the NIHR UCLH Biomedical Research Centre and the EuroPOND project— This project has received funding from the European Union's Horizon 2020 research and innovation programme under grant agreement No. 666992. AM, CZ, TW, and JM-M acknowledge funding from the Wellcome Trust under grant number WT102845/Z/13/Z.

References

1. Goriounova, N.A., Mansvelder, H.D.: Genes, cells and brain areas of intelligence. Front. Hum. Neurosci. **13**, 44 (2019). https://doi.org/10.3389/fnhum.2019.00044
2. Deary, I.J., Penke, L., Johnson, W.: The neuroscience of human intelligence differences. Nat. Rev. Neurosci. **11**(3), 201–211 (2010). https://doi.org/10.1038/nrn2793
3. McCall, R.B.: Childhood IQ's as predictors of adult educational and occupational status. Science **197**(4302), 482–483 (1977). https://doi.org/10.1126/science.197.4302.482
4. Gottfredson, L.S.: Why g matters: the complexity of everyday life. Intelligence **24**(1), 79–132 (1997). https://doi.org/10.1016/S0160-2896(97)90014-3
5. Deary, I.J., Strand, S., Smith, P., Fernandes, C.: Intelligence and educational achievement. Intelligence **35**(1), 13–21 (2007). https://doi.org/10.1016/j.intell.2006.02.001
6. Johnson, W., McGue, M., Iacono, W.G.: Genetic and environmental influences on academic achievement trajectories during adolescence. Dev. Psychol. **42**(3), 514–32 (2006). https://doi.org/10.1037/0012-1649.42.3.514
7. Batty, G.D., Deary, I.J., Gottfredson, L.S.: Premorbid (early life) IQ and later mortality risk: systematic review. Ann. Epidemiol. **17**(4), 278–288 (2007). https://doi.org/10.1016/j.annepidem.2006.07.010
8. Batty, G.D., et al.: IQ in early adulthood and mortality by middle age. Epidemiology **20**(1), 100–109 (2008). https://doi.org/10.1097/ede.0b013e31818ba076
9. Lam, N.H., et al.: Effects of Altered Excitation-Inhibition Balance on Decision Making in a Cortical Circuit Model. bioRxiv, p. 100347 (2017). https://doi.org/10.1101/100347
10. Deary, I.J., Pattie, A., Starr, J.M.: The stability of intelligence from age 11 to age 90 years: the lothian birth cohort of 1921. Psychol. Sci. **24**(12), 2361–2368 (2013). https://doi.org/10.1177/0956797613486487
11. Fors, S., Torssander, J., Almquist, Y.B.: Is childhood intelligence associated with coexisting disadvantages in adulthood? Evidence from a Swedish cohort study. Adv. Life Course Res. **38**, 12–21 (2018). https://doi.org/10.1016/J.ALCR.2018.10.005

12. MacLullich, A.M.J., Ferguson, K.J., Deary, I.J., Seckl, J.R., Starr, J.M., Wardlaw, J.M.: Intracranial capacity and brain volumes are associated with cognition in healthy elderly men. Neurology **59**(2), 169–174 (2002). https://doi.org/10.1212/WNL.59.2.169

13. McDaniel, M.A.: Big-brained people are smarter: a meta-analysis of the relationship between in vivo brain volume and intelligence. Intelligence **33**(4), 337–346 (2005). https://doi.org/10.1016/j.intell.2004.11.005

14. Rushton, J.P., Ankney, C.D.: Whole brain size and general mental ability: a review. Int. J. Neurosci. **119**(5), 692–732 (2009). https://doi.org/10.1080/00207450802325843

15. Andreasen, N.C., et al.: Intelligence and brain structure in normal individuals. Am. J. Psychiatry **150**(1), 130–4 (1993). https://doi.org/10.1176/ajp.150.1.130

16. Narr, K.L., et al.: Relationships between IQ and regional cortical gray matter thickness in healthy adults. Cereb. Cortex **17**(9), 2163–2171 (2007). https://doi.org/10.1093/cercor/bhl125

17. Karama, S., et al.: Cortical thickness correlates of specific cognitive performance accounted for by the general factor of intelligence in healthy children aged 6 to 18. NeuroImage **55**(4), 1443–1453 (2011). https://doi.org/10.1016/j.neuroimage.2011.01.016

18. Jung, R.E., Haier, R.J.: The parieto-frontal integration theory (P-FIT) of intelligence: converging neuroimaging evidence. Behav. Brain Sci. **30**(2), 135–154 discussion 154–187 (2007). https://doi.org/10.1017/S0140525X07001185

19. Gläscher, J., et al.: Lesion mapping of cognitive abilities linked to intelligence. Neuron **61**(5), 681–91 (2009). https://doi.org/10.1016/j.neuron.2009.01.026

20. Woolgar, A., et al.: Fluid intelligence loss linked to restricted regions of damage within frontal and parietal cortex. Proc. Nat. Acad. Sci. **107**(33), 14899–14902 (2010). https://doi.org/10.1073/pnas.1007928107

21. Oxtoby, N.P., Alexander, D.C., for the EuroPOND consortium: Imaging plus X: multimodal models of neurodegenerative disease. Curr. Opin. Neurol. **30**(4), 371–379 (2017). https://doi.org/10.1097/WCO.0000000000000460

22. Schrouff, J., et al.: PRoNTo: pattern recognition for neuroimaging toolbox. Neuroinformatics **11**(3), 319–37 (2013). https://doi.org/10.1007/s12021-013-9178-1

23. Schrouff, J., Monteiro, J.M., Portugal, L., Rosa, M.J., Phillips, C., Mourão-Miranda, J.: Embedding anatomical or functional knowledge in whole-brain multiple kernel learning models. Neuroinformatics **16**(1), 117–143 (2018). https://doi.org/10.1007/s12021-017-9347-8

24. Blumberg, S.B., Tanno, R., Kokkinos, I., Alexander, D.C.: Deeper image quality transfer: training low-memory neural networks for 3D images. In: Frangi, A.F., Schnabel, J.A., Davatzikos, C., Alberola-López, C., Fichtinger, G. (eds.) MICCAI 2018. LNCS, vol. 11070, pp. 118–125. Springer, Cham (2018). https://doi.org/10.1007/978-3-030-00928-1_14

25. Casey, B.J., et al.: The adolescent brain cognitive development (ABCD) study: imaging acquisition across 21 sites. Dev. Cogn. Neurosci. **32**, 43–54 (2018). https://doi.org/10.1016/j.dcn.2018.03.001

26. Hagler, D.J., et al.: Image processing and analysis methods for the Adolescent Brain Cognitive Development Study. bioRxiv, p. 457739 (2018). https://doi.org/10.1101/457739

27. Pfefferbaum, A., et al.: Altered brain developmental trajectories in adolescents after initiating drinking. Am. J. Psychiatry **175**(4), 370–380 (2018). https://doi.org/10.1176/appi.ajp.2017.17040469

28. Rohlfing, T., Zahr, N.M., Sullivan, E.V., Pfefferbaum, A.: The SRI24 multichannel atlas of normal adult human brain structure. Hum. Brain Mapp. **31**(5), 798–819 (2010). https://doi.org/10.1002/hbm.20906

29. Akshoomoff, N., et al.: VIII. NIH toolbox cognition battery (CB): composite scores of crystallized, fluid, and overall cognition. Monogr. Soc. Res. Child Dev. **78**(4), 119–132 (2013). https://doi.org/10.1111/mono.12038

30. Monté-Rubio, G.C., Falcón, C., Pomarol-Clotet, E., Ashburner, J.: A comparison of various MRI feature types for characterizing whole brain anatomical differences using linear pattern recognition methods. NeuroImage **178**, 753–768 (2018). https://doi.org/10.1016/j.neuroimage.2018.05.065

31. Ashburner, J.: A fast diffeomorphic image registration algorithm. NeuroImage **38**(1), 95–113 (2007). https://doi.org/10.1016/J.NEUROIMAGE.2007.07.007

32. Blaiotta, C., Freund, P., Cardoso, M.J., Ashburner, J.: Generative diffeomorphic modelling of large MRI data sets for probabilistic template construction. NeuroImage **166**, 117–134 (2018). https://doi.org/10.1016/j.neuroimage.2017.10.060

33. Rakotomamonjy, A., Bach, F.R., Canu, S., Grandvalet, Y.: SimpleMKL. J. Mach. Learn. Res. **9**, 2491–2521 (2008). http://www.jmlr.org/papers/v9/rakotomamonjy08a.html

34. Shawe-Taylor, J., Cristianini, N.: Kernel Methods for Pattern Analysis. Cambridge University Press, Cambridge (2004)

35. Rasmussen, C.E., Williams, C.K.I.: Gaussian Processes for Machine Learning. The MIT Press, Cambridge (2006). http://www.GaussianProcess.org/gpml

36. Tipping, M.E.: Sparse Bayesian learning and the relevance vector machine. J. Mach. Learn. Res. **1**, 211–244 (2001). http://www.jmlr.org/papers/v1/tipping01a.html

37. Sturmfels, P., Rutherford, S., Angstadt, M., Peterson, M., Sripada, C., Wiens, J.: A domain guided CNN architecture for predicting age from structural brain images. In: Doshi-Velez, F., et al. (eds.) Proceedings of the 3rd Machine Learning for Healthcare Conference. Proceedings of Machine Learning Research, Palo Alto, California, vol. 85, pp. 295–311. PMLR (2018). http://proceedings.mlr.press/v85/sturmfels18a.html

38. Payan, A., Montana, G.: Predicting Alzheimer's disease: a neuroimaging study with 3D convolutional neural networks. arXiv e-prints arXiv:1502.02506, February 2015. Preprint

39. Milletari, F., et al.: Hough-CNN: Deep Learning for Segmentation of Deep Brain Regions in MRI and Ultrasound. arXiv e-prints arXiv:1601.07014 (2016)

40. Rao, A., Monteiro, J.M., Mourao-Miranda, J.: Alzheimer's disease initiative: predictive modelling using neuroimaging data in the presence of confounds. NeuroImage **150**, 23–49 (2017). https://doi.org/10.1016/j.neuroimage.2017.01.066

41. Varoquaux, G.: Cross-validation failure: small sample sizes lead to large error bars. NeuroImage **180**, 68–77 (2018). https://doi.org/10.1016/j.neuroimage.2017.06.061

Predicting Fluid Intelligence Using Anatomical Measures Within Functionally Defined Brain Networks

Jeffrey N. Chiang[1](✉), Nicco Reggente[2](✉), John Dell'Italia[1](✉),
Zhong Sheng Zheng[3](✉), and Evan S. Lutkenhoff[1](✉)

[1] Department of Psychology, University of California, Los Angeles,
Los Angeles, CA, USA
njchiang@g.ucla.edu, {johndellitalia,lutkenhoff}@ucla.edu
[2] Tiny Blue Dot Foundation, Santa Monica, CA, USA
nicco@tinybluedotfoundation.org
[3] Casa Colina Research Institute, Pomona, CA, USA
azszheng@gmail.com

Abstract. The ABCD Neurocognitive Prediction Challenge (ABCD-NP-Challenge 2019) made available T1-weighted structural scans for children alongside their fluid intelligence scores. The goal of the challenge was to use this anatomical brain data to train a model that could be successful in predicting fluid intelligence scores from held-out T1-weighted structural scans taken of other children. Functional magnetic resonance imaging (fMRI) has been moderately successful at identifying neural correlates of cognitive functioning, including intelligence. This study sought to leverage anatomical metrics within functionally defined regions, convolutional neural networks, and regression models to predict fluid intelligence. The proposed model performed competitively on the ABCD-NP-Challenge, and significantly outperformed a non deep-learning approach for behavior prediction based on the LASSO.

Keywords: Intelligence · Network analysis · Convolutional neural network

1 Introduction

Functional connectivity within cortical networks has shown to be informative enough to permit predictions of individual differences in many different domains: cognitive control/intelligence [5], visual tasks [1], brain maturity [7], and outcome response to cognitive behavioral therapy for patients with OCD [15]. Given the predictive efficacy of the functional profiles of these networks, coupled with a desire to leverage neural metrics that are more easily accessible, we sought to identify the degree to which static anatomical measures contained information pertaining to individual differences in intelligence. Of particular interest was the Cingulo-Opercular Network (CON), which has been shown to support a

K. M. Pohl et al. (Eds.): ABCD-NP 2019, LNCS 11791, pp. 143–149, 2019.
https://doi.org/10.1007/978-3-030-31901-4_17

basic domain-independent and externally directed "task mode" (e.g., [7]). Such a function suggests that the CON could theoretically be the driving factor for supporting fluid intelligence given the "domain-independent" nature of such a metric.

When using a least absolute shrinkage and selection operator (LASSO) regression model, [15] were successful in using pre-treatment neural metrics to predict response to cognitive behavioral therapy in patients with obsessive compulsive disorder. As such, we first sought to test the effectiveness of a LASSO when predicting intelligence.

In parallel, convolutional neural networks (CNNs) have become the method of choice for image-based machine learning applications, including brain-based MRI analysis [2]. Much of their success has been attributed to their ability to discover latent features without an *a priori* specification of the latent structure. However, a key limitation of neural network-based methods is that they require large amounts of training data, preventing their use in certain clinical and diagnostic applications, instead being used in structured image processing tasks such as brain segmentation. However, with the release of the ABCD-NP-Challenge dataset, we are able to apply these methods, leveraging them with neurobiologically relevant measures to predict the fluid intelligence of subjects given their brain anatomy.

In this report, we leverage and compare recent advances in automated biomedical image analysis (i.e. CNNs and LASSO) and neurobiologically relevant measures to refine estimates of fluid intelligence.

2   Methods

The ABCD-NP-Challenge research consortium [18] provided anatomical and demographic data for children aged 9–10 years (n = 8669) as part of the MICCAI ABCD-NP-Challenge. T1-weighted MRI data, along with fluid intelligence scores were released in training (n = 3739), testing (n = 4515), and validation (n = 415) sets.

In order to encourage participants to utilize image-based features, the challenge organizers regressed out common behavioral and demographic factors out of the measured fluid intelligence scores. These residualized scores served as the target for the model to predict.

2.1   ROI Creation

The brain was segmented into 390 regions-of-interest (ROIs) using three different MR-atlas based approaches (see Fig. 1 for an example subject with the three atlases). First, 15 subcortical regions (brainstem, bilateral nucleus accumbens, amygdala, caudate, hippocampus, globus pallidus, putamen, and thalamus) were segmented per subject using FSL FIRST [13]. Second, 109 regions were derived from the SRI24 atlas [16] with registrations between the MNI152 T1 (2 mm) and SRI24 templates were accomplished with FLIRT. These regions

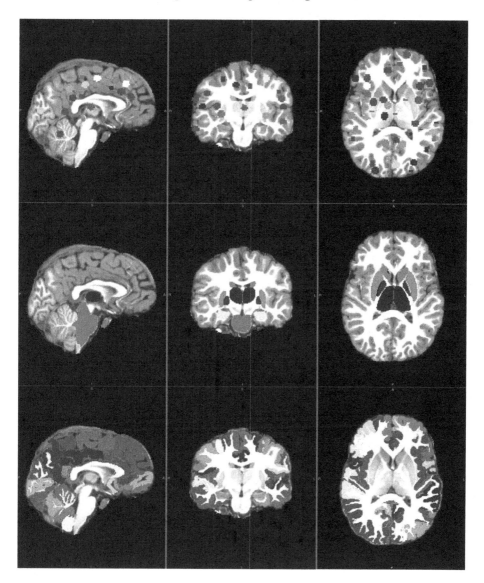

Fig. 1. ROIs leveraged in the LASSO analysis from 3 atlases: Petersen (Row 1), FSL FIRST (Row 2), and SRI24 (Row 3).

were mainly cortical and did not include any of the subcortical ROIs from the FSL FIRST method. Third, 5 mm spherical, network-defined regions of interest (ROIs) were derived from [14] (also referenced as the Petersen atlas) for the Visual, Fronto Parietal, Somatosensory, Motor, Default Mode Network, and Cingulo Opercular Network. An additional two theory-driven networks (Resonance and Control) were created based by selecting ROIs from [14] that overlapped with trait-empathy activation maps from [4]. An additional pair of 5 mm ROI's centered on left (x = −22 mm, y = −6 mm, z = −14 mm) and right (x = 22 mm, y = −6 mm, z = −14 mm) amygdala were added to the resonance network, due to their relevance for emotional forms of self-other resonance [e.g., 3,4]. Each of the 390 ROIs were affinely registered from the atlas spaces to each individuals' brain-extracted T1-weighted MRI image using FSL FLIRT [8,9]. Then, in individual subject space, the total volume, mean signal intensity, intensity standard deviation, and entropy were calculated per ROI using FSL's fslstats tool [17].

2.2 Intelligence Prediction via Least Absolute Shrinkage and Selection Operator LASSO

We leveraged a least absolute shrinkage and selection operator (LASSO) regression model built on the feature sets (total volume, mean signal intensity, and entropy within each of the nodes making up one cortical network from the eight networks outlined above) available for each subject in the training set where the outcome variable was the subject's residualized fluid intelligence score provided by ABCD-NP-Challenge. The model's intercept term and outcome beta values were then used as coefficients for each left-out subject in the testing set, obtaining a predicted subscale measure for that individual. The regularization parameter for the LASSO was optimized previously on an N-1 cross-validation within the test set that maximized the Pearson correlation between actual and predicted IRI fluid intelligence scores [15]. This process was repeated for each of the eight networks, yielding an array of network-specific accuracy measures.

2.3 Intelligence Prediction via Convolutional Neural Network

Another model was designed to both capture novel, 3d volumetric features from the raw T1-weighted image as well as use anatomically relevant information derived from the SRI24 and Petersen atlases described above. Thus the model consisted of two processing streams, a convolutional neural network operating on the raw T1-weighted image, and a fully connected network operating on the derived ROI features.

The convolutional network was inspired by the false-positive reduction architecture described by [6], which was built on 3d convolutional blocks and would capture any volumetric features missed by the manual feature-extraction. The network was built of 6 convolutional blocks, consisting of a 3d convolution + batch normalization + leaky ReLU, each with 3d max pooling with stride 2 in between in order to reduce the volume by a factor of two. The first convolutional block had a filter size of (15, 15, 15) with 16 channels, and following blocks used a filter size of (3, 3, 3) with 16, 32, 64, 64, and 64 channels respectively. After the final convolutional block, a Global Average Pooling was computed in order to make the resulting feature map compatible with ROI derived features. Due to GPU memory constraints, the raw T1-weighted image was first resampled to have $2 \times 2 \times 2$ mm voxel size, and further cropped to a size (79, 91, 79), taking advantage of the fact that all volumes were first spatially aligned before submitting to analysis.

The manually extracted features were passed through a fully connected network. The volume and entropy features for the Petersen ROIs computed above were concatenated to form a 1×780 vector, which was passed through a fully connected network of 128, then 64 hidden units, each with ReLU activation.

The final MSE prediction was derived by adding the resulting outputs of the CNN (1×64) and fully connected network (1×64) together, passing them through a fully connected layer with 128 hidden units, then arriving at a single output unit optimized using mean squared error loss against the training data. The model was trained using the Adam optimizer [10] with the learning rate set to 1e-3 and run for 10000 iterations, then lowered to 1e-4. All training was computed on an NVIDIA RTX 2070 gpu and implemented using Tensorflow 1.13.

3   Results

The LASSO method, a non deep learning approach, was trained using the collection of volume, mean signal intensity, intensity standard deviation, and entropy within ROIs composing each of the networks and atlases described in the methods. None of the examined collections of ROIs yielded a significant predictive power, with the best performing feature set (Default Mode Network) yielding an MSE of 81 on the ABCD-NP-Challenge validation set.

The proposed model, a convolutional neural network using the raw T1 image plus volume and entropy features from above, received a MSE of 71.89 on the ABCD-NP-Challenge validation set, and a MSE of 95.38 on the test set (the winning score was 92.13). This model outperformed the LASSO method with theory-driven feature extraction by a large margin.

4   Discussion

The goal of the ABCD-NP-Challenge was to predict childrens' fluid intelligence scores based on brain anatomy, derived from T1-weighted images, after accounting for demographic factors. This provided an opportunity to bridge a gap between recent advances in machine learning applications, specifically 3d convolutional neural networks, with established brain mapping literature. Using this approach, our model achieved a competitive result on the provided validation dataset. Additional regression methods, i.e. LASSO, were also implemented but failed to reach statistical significance. Multiple measures per ROI were used including total volume, mean signal intensity, signal intensity standard deviation, and entropy.

While the target residualized fluid intelligence scores were designed to encourage competition participants to utilize volumetric and anatomical information, it is possible (and likely) that some effect of brain anatomy, which is correlated with development [12], was removed from the target scores during this procedure, preventing higher predictive power.

Previous research has investigated the anatomical and functional mechanisms underlying fluid intelligence but both, particularly the former, have remained elusive. Our model shows the potential for combining a priori anatomical and functional knowledge with modern machine learning approaches for analysis. Further research will need to be conducted to determine which of the functionally defined networks from the introduction are most informative in predicting fluid intelligence – a finding that could potentially be elucidated within the construct of a recursive feature elimination methodology [11]. Additional analyses could also include multiple spatial resolutions of the data into the CNN to leverage the voxel level, ROI level, and network level features into a single model, which would optimize all the spatial scales together rather than independently testing each spatial resolutions. This would enable us to utilize, in parallel, insights from systems spanning multiple spatial scales (e.g. from neural columns to networks).

References

1. Baldassarre, A., Lewis, C.M., Committeri, G., Snyder, A.Z., Romani, G.L., Corbetta, M.: Individual variability in functional connectivity predicts performance of a perceptual task. Proc. Natl. Acad. Sci. **109**(9), 3516–3521 (2012)
2. Bernal, J., et al.: Deep convolutional neural networks for brain image analysis on magnetic resonance imaging: a review. Artif. Intell. Med. **95**, 64–81 (2018)
3. Carr, L., Iacoboni, M., Dubeau, M.C., Mazziotta, J.C., Lenzi, G.L.: Neural mechanisms of empathy in humans: a relay from neural systems for imitation to limbic areas. Proc. Natl. Acad. Sci. **100**(9), 5497–5502 (2003)
4. Christov-Moore, L., Iacoboni, M.: Self-other resonance, its control and prosocial inclinations: brain-behavior relationships. Hum. Brain Mapp. **37**(4), 1544–1558 (2016)
5. Cole, M.W., Yarkoni, T., Repovš, G., Anticevic, A., Braver, T.S.: Global connectivity of prefrontal cortex predicts cognitive control and intelligence. J. Neurosci. **32**(26), 8988–8999 (2012)
6. Ding, J., Li, A., Hu, Z., Wang, L.: Accurate pulmonary nodule detection in computed tomography images using deep convolutional neural networks. In: Descoteaux, M., Maier-Hein, L., Franz, A., Jannin, P., Collins, D.L., Duchesne, S. (eds.) MICCAI 2017. LNCS, vol. 10435, pp. 559–567. Springer, Cham (2017). https://doi.org/10.1007/978-3-319-66179-7_64
7. Dosenbach, N.U., et al.: Prediction of individual brain maturity using fMRI. Science **329**(5997), 1358–1361 (2010)
8. Jenkinson, M., Bannister, P., Brady, M., Smith, S.: Improved optimization for the robust and accurate linear registration and motion correction of brain images. Neuroimage **17**(2), 825–841 (2002)
9. Jenkinson, M., Smith, S.: A global optimisation method for robust affine registration of brain images. Med. Image Anal. **5**(2), 143–156 (2001)
10. Kingma, D.P., Ba, J.: Adam: a method for stochastic optimization. arXiv preprint arXiv:1412.6980 (2014)
11. Louw, N., Steel, S.: Variable selection in Kernel Fisher discriminant analysis by means of recursive feature elimination. Comput. Stat. Data Anal. **51**(3), 2043–2055 (2006)

12. Noble, K.G., Houston, S.M., Kan, E., Sowell, E.R.: Neural correlates of socioeconomic status in the developing human brain. Dev. Sci. **15**(4), 516–527 (2012)
13. Patenaude, B., Smith, S.M., Kennedy, D.N., Jenkinson, M.: A bayesian model of shape and appearance for subcortical brain segmentation. Neuroimage **56**(3), 907–922 (2011)
14. Power, J.D., et al.: Functional network organization of the human brain. Neuron **72**(4), 665–678 (2011)
15. Reggente, N., et al.: Multivariate resting-state functional connectivity predicts response to cognitive behavioral therapy in obsessive-compulsive disorder. Proc. Natl. Acad. Sci. **115**(9), 2222–2227 (2018)
16. Rohlfing, T., Zahr, N.M., Sullivan, E.V., Pfefferbaum, A.: The SRI24 multichannel atlas of normal adult human brain structure. Hum. Brain Mapp. **31**(5), 798–819 (2010)
17. Smith, S.M., et al.: Advances in functional and structural MR image analysis and implementation as FSL. Neuroimage **23**, S208–S219 (2004)
18. Volkow, N.D., et al.: The conception of the ABCD study: from substance use to a broad NIH collaboration. Dev. Cogn. Neurosci. **32**, 4–7 (2018)

Sex Differences in Predicting Fluid Intelligence of Adolescent Brain from T1-Weighted MRIs

Sara Ranjbar$^{(\boxtimes)}$ ⓘ, Kyle W. Singleton ⓘ, Lee Curtin ⓘ,
Susan Christine Massey ⓘ, Andrea Hawkins-Daarud ⓘ,
Pamela R. Jackson ⓘ, and Kristin R. Swanson ⓘ

Mathematical NeuroOncology Lab, Precision Neurotherapeutics Innovation
Program, Department of Neurological Surgery, Mayo Clinic, Phoenix, AZ, USA
ranjbar.sara@mayo.edu

Abstract. Fluid intelligence (Gf) has been defined as the ability to reason and solve previously unseen problems. Links to Gf have been found in magnetic resonance imaging (MRI) sequences such as functional MRI and diffusion tensor imaging. As part of the Adolescent Brain Cognitive Development Neurocognitive Prediction Challenge 2019, we sought to predict Gf in children aged 9–10 from T1-weighted (T1 W) MRIs. The data included atlas–aligned volumetric T1 W images, atlas–defined segmented regions, age, and sex for 3739 subjects used for training and internal validation and 415 subjects used for external validation. We trained sex-specific convolutional neural net (CNN) and random forest models to predict Gf. For the convolutional model, skull-stripped volumetric T1 W images aligned to the SRI24 brain atlas were used for training. Volumes of segmented atlas regions along with each subject's age were used to train the random forest regressor models. Performance was measured using the mean squared error (MSE) of the predictions. Random forest models achieved lower MSEs than CNNs. Further, the external validation data had a better MSE for females than males (60.68 vs. 80.74), with a combined MSE of 70.83. Our results suggest that predictive models of Gf from volumetric T1 W MRI features alone may perform better when trained separately on male and female data. However, the performance of our models indicates that more information is necessary beyond the available data to make accurate predictions of Gf.

Keywords: Fluid intelligence · Sex differences · Deep learning

1 Introduction

Fluid intelligence (Gf) is the ability to reason and solve previously unseen problems [1]. It is highly associated with general intelligence, more so than the other intelligence subtypes [2] and has been linked to academic performance [3]. It is generally theorized that Gf is highly dependent upon biological processes and as such is more independent from previous learning than other types of intelligence, such as crystalized intelligence [4]. Gf is known to increase throughout childhood, peak around early adulthood and then decrease throughout adult life [4–7]. Fluid intelligence is typically measured using a battery of tests evaluating aspects of memory and pattern recognition [4].

© Springer Nature Switzerland AG 2019
K. M. Pohl et al. (Eds.): ABCD-NP 2019, LNCS 11791, pp. 150–157, 2019.
https://doi.org/10.1007/978-3-030-31901-4_18

Magnetic resonance imaging (MRI) is a powerful tool for non-invasively visualization of the body's tissues. Structural MRI, including T1-weighted (T1 W) and T2-weighted scans, exhibit excellent contrast for non-invasively discriminating many brain structures. Advanced and quantitative MRI can provide more than just structural information, including brain activity as detected by functional MRI (fMRI) and white matter tractography from diffusion tensor imaging (DTI). Previous studies have shown that fluid intelligence can be predicted by imaging, in particular fMRI [2, 8] and white matter tract diffusion tensors from DTI [6]. In addition, levels of N-acetylaspartate and brain volume from MRI spectroscopy were shown to be associated with aspects of Gf, but not Gf itself [9]. To our knowledge, fluid intelligence has not been previously linked to T1 W imaging.

Discovering connections between cognitive traits and non-invasive biomedical imaging could prove to be important for further understanding the neural underpinnings of cognitive development. Recently deep learning models trained on brain MRI data have shown promising results in the diagnosis of Alzheimer's disease [10], prediction of age [11, 12] and classification of overall survival in brain tumor patients [8]. In this work we compared the performance of deep learning techniques with that of classic machine learning techniques trained on hand-crafted features in the prediction of Gf from MRI-based features. We also opted to train separate models for male and female cohorts. The potential impact of biological sex differences is a recommended consideration for neuroscience research [13]. Overall, males have been reported to have larger brains, and in young subjects who are still developing, sex differences have been reported in the growth and maturation of various brain structures [14, 15]. Consequently, sex differences in the developmental state of brain structures may affect MRI features, including the size and texture of various brain regions. It is not yet clear if these factors are related to Gf. By conducting our analysis in a sex–specific manner, we sought to reduce any erroneous association between the development of brain structures and Gf.

2 Material

Data used in this work was acquired by the Adolescent Brain Cognitive Development (ABCD) study (abcdstudy.org) and access was provided to participants as part of the ABCD Challenge (sibis.sri.com/abcd-np-challenge). The ABCD study is the largest long-term study of brain development and child health in the United States. The overall challenge cohort included data from 4154 children ages 9–10. Extensive information about ABCD data can be found in the Data supplement of [16]. Data was split into 3739 subjects in the training set (47.4% of which were females) and 415 (49.3% females) in the external validation set. The provided skull-stripped T1 W MRIs underwent pre-processing for conversion to the NifTI format, noise removal, and field inhomogeneity correction as described in [17]. Further, all volumetric MRIs were standardized to a 240 × 240 × 240 dimensionality with voxel resolution of 1 mm in x, y, and z directions. The images were also affinely aligned to the SRI 24 atlas [18] and segmented into cortical and subcortical structures (e.g. parcellated gray matter, white matter, and cerebral fluid) according to the atlas. Volumetric scores of 122 segmented brain regions, along with participants' sex and age at interview in months were provided to participants in the format of a csv file. Accompanying fluid intelligence scores were assessed using a variety of tests as detailed in [4]. Scores were normalized using a linear regression model trained on factors such as brain volume, data collection site, age at baseline, sex at birth, race/ethnicity, highest parental education, parental income, and parental marital status. Demographic confounding factors such as sex and age were removed from the scores by the ABCD study. Figure 1 shows the distribution of fluid intelligence across the training and validation sets for males and females.

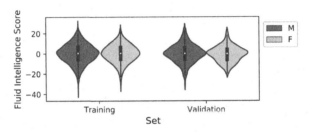

Fig. 1. Distribution of fluid intelligence scores across training and validation sets for males and females.

3   Methods

We conducted two separate experiments to predict fluid intelligence scores: in the first, we used skull-stripped T1 W images as the input to train 3D convolutional neural nets to predict Gf scores. In the second, we trained random forest regression models using the volumetric features from image-atlas alignment along with the age and sex of subjects. For both experiments, we trained models on sex–specific cohorts and predicted the Gf of sex-equivalent subjects in the validation set. Male and female predictions were used to compute sex-specific MSE scores, and then a total population MSE was calculated using the merged set of sex-specific predictions. For a comparative baseline performance, we created a predictor with zero rule algorithm using the mean Gf in the training cohort and measured the model accuracy in predicting Gf for the validation set.

3.1   Convolutional Neural Networks

We used residual convolutional neural networks in the first experiment. The residual architecture (ResNet) has previously been described in [11]. Nets were trained on a Nvidia TITAN V GPU using Keras with TensorFlow backend running through Nvidia docker, Python 3, and TensorFlow 1.12.0. To adjust for the computational capabilities of our GPU, we resized the images to $128 \times 128 \times 128$ voxels and used a stride of 2 in the initial convolutional layer. Several areas of the brain, such as the caudate and putamen, have previously been reported to have links to fluid intelligence [19]. As a result, we repeated training with the ResNet using only slices that included the caudate and putamen (i.e., slices 55–75) resulting in image dimensionality of $128 \times 128 \times 20$. Training was performed using the RMSprop optimization algorithm for learning the weights, a batch size of 32, an MSE loss function, and 100 epochs. Learning rate was initially set to 0.001 and was reduced by a factor of 0.1 at validation loss plateau.

As mentioned before, separate models were generated for male and female cohorts. A total of 1966 MRIs were available for the male model and 1773 MRIs for the female model. Prior to training, the sex-specific training sets were further split into training and internal validation sets using an 80:20 ratio. We did not perform data augmentation in this work.

3.2 Random Forests

In the second experiment we trained random forest regression models using volumetric and demographic features described in Sect. 2. We used the scikit-learn [20] package and Python 3.6.8 for this experiment. Hyperparameter selection was performed using grid search for maximum depth (range of 2–6) and the number of estimators in the random forest (100–500, 750, and 1000 trees). The remaining hyperparameters were left at default values in scikit-learn. The same set of 1966 male and 1773 female subjects were used in these sex-specific models.

4 Results

Males and females shared a median age of 120 months in training and validation sets. The ratio of males to females was 1.11:1 in the training set and 1.02:1 in the validation set for a total of 1.10:1 for the entire cohort.

4.1 CNN Analysis Results

Overall performance of the CNN attempts was poor. Sex-specific ResNet using the complete T1 W MRIs had better total performance than the caudate-putamen T1 W slices alone. However, MSE was much lower in female cases in the caudate-putamen case, indicating possible influence of sex on those regions. None of the CNN models evaluated in our analysis was able to obtain a lower MSE than the baseline predictor.

4.2 Random Forest Analysis Results

Table 1 shows the top-10 most important features to the male and female random forest models. Pons white matter volume was most important for both sexes, which may highlight the important role of sensory processing carried out by the pons [21]. Yet, there are notable differences in the selected features between the two models. In particular, hippocampus volumes (both left and right) were important among males but not females. Since the hippocampus is generally considered to play a pivotal role in the consolidation of short-term to long-term memory through detection of new stimuli [22], and is also active in navigation and spatial memory [23, 24], it is striking that it only appeared to be of considerable importance in one sex. Overall, random forest models utilizing demographic and brain region volume data had the best performance in our analysis with a smaller MSE than the baseline predictor.

Table 1. Top 10 important features along with their importance scores for male and female random forest models. For further information on what these features represent the avid reader is referred to the supplement of [16].

Male		Female	
Feature	Importance	Feature	Importance
ponsWM	0.042	ponsWM	0.0523
hippocampus_L_GM	0.041	cerebelum8_R_VOL	0.0391
cerebelum6_R_VOL	0.0327	temporalinf_L_GM	0.0282
cerebelum45_R_VOL	0.0316	WM400WM400_R_WM	0.0269
hippocampus_R_GM	0.0277	fusiform_R_GM	0.0209
parietalinf_R_GM	0.0254	calcarine_R_GM	0.0207
frontalmid_L_GM	0.0189	temporalsup_L_GM	0.02
parahippocampal_L_GM	0.0181	precuneus_L_GM	0.019
amygdala_L_GM	0.0181	cuneus_R_GM	0.017
cerebelumcrus2_L_VOL	0.0175	cerebelum45_R_VOL	0.0162

4.3 Comparison of CNN and Random Forest

Table 2 compares the performance of CNN and random forest models in prediction of Gf. ResNet model MSE values were significantly higher than the baseline predictor in a 2-sided t-test (ResNet Full Brain: p-value = 0.0006; ResNet Caudate-Putamen: p-value = 7.9e-5). The difference in the MSE for random forest models from the baseline predictor was lower by 1.01, but this difference was not significantly different from the baseline predictor (p-value = 0.17). The female random forest model presented a lower MSE than the male model, similar to the trends observed for ResNet models. Figure 2 compares the distribution of predicted Gf scores all models. As evident, all models failed to predict the full range of Gf scores in the validation set and yielded predictions close to the average Gf.

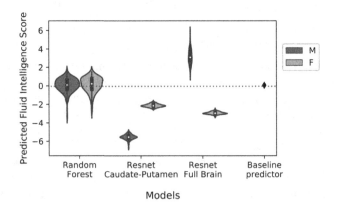

Fig. 2. Distributions of predicted fluid intelligence scores in our experiments. Dashed line shows the average fluid intelligence score in the validation set.

Table 2. Summary of the performance of models on internal/external validation sets

Classification method	Sex (Training)	External validation MSE score	
		Sex specific	Combined population
Resnet (Full Brain)	M	93.57	79.23
Resnet (Full Brain)	F	77.12	
Resnet (Caudate-Putamen)	M	111.83	87.05
Resnet (Caudate-Putamen)	F	61.67	
Random Forests	M	80.74	70.83
Random Forests	F	60.68	
Baseline predictor	–	–	71.84

5 Discussion

Fluid intelligence has been an established metric in psychology and education research since the early 70's [25]. Standard Gf assessments primarily rely on non-verbal multiple choice questionnaires [26]. While there have been a few studies investigating whether Gf scores are predictable from functional MRI, to date, there have been no studies investigating possible connections from structural T1 W MRI. As part of the ABCD challenge, we present a new line of investigation into whether anatomical attributes of an individual's brain as visualized on a T1 W MRI are predictive of Gf.

In this work, we investigated this hypothesis by assessing the performance of different machine learning models in predicting Gf. Convolutional neural network models were trained on slices presenting selected brain structures (caudate and putamen) as well as on the entire volume of T1 W MRI images. Random forest regressor was trained using the volumes of different brain regions. We focused on sex-specific versions of these models due to previously observed sex-differences in structural MRI in the growth and maturation of various brain structures in children and adolescent subjects [14, 15]. We chose to consider both the CNNs and the random forest as they have different strengths at evaluating the available data. Random forest is a powerful machine learning method capable of incorporating both categorical and numerical hand-crafted features in the decision-making process. CNNs on the other hand, do not require hand-crafted features and are able to extract imaging patterns at both small and large scales. Since none of the models presented a compelling accuracy separately, we did not merge the two into a combined model.

Given the large size of the ABCD dataset, it was somewhat surprising that CNN was outperformed by the random forest model. However, a closer look at the predictions indicated that they were all tightly clustered around one or another Gf score (Fig. 2) demonstrating that all models failed to predict the full range of intelligence scores. The random forest model marginally described the range of Gf scores and therefore achieved a lower MSE than CNNs. This strongly suggests that the relationship between Gf and imaging features seen on T1 W MRI is not predictive. It was also somewhat unexpected that the female model outperformed the male model. The large difference in MSE between the female and male models is likely attributable to the reduced number of outlier Gf scores in the female validation subcohort (Fig. 1).

However, the fact that volumes of different brain regions contributed to the two models (Table 1) may indicate sex differences in the underlying mechanisms of fluid intelligence in developing brain. Further investigation is needed to determine whether there are in fact different imaging predictors of Gf among developing young male vs. female brain.

Ultimately, T1 W MRIs and volumetric features did not provide a compelling accuracy in prediction of fluid intelligence scores in our analyses. While it is possible that our approaches were not sophisticated enough to detect the predictive relationship, the similarity of all approaches to a baseline zero rule algorithm using the training set mean suggests that the characteristics of physical brain structures as identifiable on T1 W MRI do not explain Gf, despite being individualized. Other MRI sequences that have been reported to link with Gf, such as fMRI [2, 8] and DTI [6], contain more functional information than T1 W MRI, suggesting that additional information is required to capture a relationship between structural imaging and fluid intelligence. It should also be noted that the complexity of this proposed problem could be confounded by the Gf scoring system.

6   Conclusion

In this work we experimented with predicting fluid intelligence of adolescent brain using MRI and a set of machine learning techniques. Overall, the performance of the best model was not significantly superior to a baseline predictor using the average fluid intelligence score. We associate the uninspiring predictive performance of the models to the insufficiency of structural MRI in explaining the complexity behind fluid intelligence mechanism in developing brain.

Acknowledgements. The authors would like to thank the Challenge Organizers and ABCD Study Researchers for the opportunity to participate and utilize their data. We also thank Kevin Flores, Erica Rutter, and John Nardini for many helpful discussions. Further, we acknowledge the following funding sources: James S. McDonnell Foundation, U54CA210180, U54CA193489, 3U54CA193489-04S3, and U01CA220378.

References

1. Merrifield, P.R., Cattell, R.B.: Abilities: their structure, growth, and action (1975). https://doi.org/10.2307/1162752
2. Colom, R., et al.: Gray matter correlates of fluid, crystallized, and spatial intelligence: testing the P-FIT model (2009). https://doi.org/10.1016/j.intell.2008.07.007
3. Chamorro-Premuzic, T., Furnham, A.: Personality, intelligence and approaches to learning as predictors of academic performance (2008). https://doi.org/10.1016/j.paid.2008.01.003
4. Akshoomoff, N., et al.: VIII. NIH Toolbox Cognition Battery (Cb): Composite Scores of Crystallized, Fluid, and Overall Cognition (2013). https://doi.org/10.1111/mono.12038
5. Horn, J.L., Cattell, R.B.: Age differences in fluid and crystallized intelligence (1967). https://doi.org/10.1016/0001-6918(67)90011-x

6. Kievit, R.A., Davis, S.W., Griffiths, J., Correia, M.M., Cam-Can, Henson, R.N.: A watershed model of individual differences in fluid intelligence. Neuropsychologia **91**, 186–198 (2016)
7. Fry, A.F., Hale, S.: Relationships among processing speed, working memory, and fluid intelligence in children. Biol. Psychol. **54**, 1–34 (2000)
8. Finn, E.S., et al.: Functional connectome fingerprinting: identifying individuals using patterns of brain connectivity. Nat. Neurosci. **18**, 1664–1671 (2015)
9. Paul, E.J., et al.: Dissociable brain biomarkers of fluid intelligence. Neuroimage **137**, 201–211 (2016)
10. Cole, J.H., et al.: Predicting brain age with deep learning from raw imaging data results in a reliable and heritable biomarker (2017). https://doi.org/10.1016/j.neuroimage.2017.07.059
11. Yang, C., Rangarajan, A., Ranka, S.: Visual explanations from deep 3D convolutional neural networks for alzheimer's disease classification. In: AMIA Annual Symposium Proceedings, pp. 1571–1580 (2018)
12. Shi, J., Zheng, X., Li, Y., Zhang, Q., Ying, S.: Multimodal Neuroimaging Feature Learning With Multimodal Stacked Deep Polynomial Networks for Diagnosis of Alzheimer's Disease. IEEE J Biomed Health Inform. **22**, 173–183 (2018)
13. Shansky, R.M., Woolley, C.S.: Considering sex as a biological variable will be valuable for neuroscience research. J. Neurosci. **36**, 11817–11822 (2016)
14. Sowell, E.R., Trauner, D.A., Gamst, A., Jernigan, T.L.: Development of cortical and subcortical brain structures in childhood and adolescence: a structural MRI study. Dev. Med. Child Neurol. **44**, 4–16 (2002)
15. Giedd, J.N., Rapoport, J.L.: Structural MRI of pediatric brain development: what have we learned and where are we going? Neuron **67**, 728–734 (2010)
16. Pfefferbaum, A., et al.: Altered brain developmental trajectories in adolescents after initiating drinking. Am. J. Psychiatry **175**, 370–380 (2018)
17. Hagler, D.J., et al.: Image processing and analysis methods for the adolescent brain cognitive development study (2018). https://www.biorxiv.org/content/early/2018/11/04/457739
18. Rohlfing, T., Zahr, N.M., Sullivan, E.V., Pfefferbaum, A.: The SRI24 multichannel atlas of normal adult human brain structure. Hum. Brain Mapp. **31**, 798–819 (2010)
19. Burgaleta, M., et al.: Subcortical regional morphology correlates with fluid and spatial intelligence. Hum. Brain Mapp. **35**, 1957–1968 (2014)
20. Pedregosa, F., et al.: Scikit-learn: machine learning in Python. J. Mach. Learn. Res. **12**, 2825–2830 (2011)
21. Saladin, K.S.: Anatomy & Physiology: The Unity of Form and Function. McGraw-Hill Science, Engineering & Mathematics, New York (2007)
22. VanElzakker, M., Fevurly, R.D., Breindel, T., Spencer, R.L.: Environmental novelty is associated with a selective increase in Fos expression in the output elements of the hippocampal formation and the perirhinal cortex. Learn. Mem. **15**, 899–908 (2008)
23. Duarte, I.C., Ferreira, C., Marques, J., Castelo-Branco, M.: Anterior/posterior competitive deactivation/activation dichotomy in the human hippocampus as revealed by a 3D navigation task. PLoS ONE **9**, e86213 (2014)
24. Maguire, E.A., et al.: Navigation-related structural change in the hippocampi of taxi drivers. Proc. Natl. Acad. Sci. U.S.A. **97**, 4398–4403 (2000)
25. Cattell, R.B.: Abilities: Their Structure, Growth, and Action. Houghton Mifflin Harcourt (HMH), Boston (1971)
26. Raven, J.C., Court, J.H.: Manual for Raven's progressive matrices and vocabulary scales: advanced progressive matrices (1998)

Ensemble of 3D CNN Regressors with Data Fusion for Fluid Intelligence Prediction

Marina Pominova, Anna Kuzina, Ekaterina Kondrateva[✉],
Svetlana Sushchinskaya, Evgeny Burnaev, Vyacheslav Yarkin,
and Maxim Sharaev

Skolkovo Institute of Science and Technology, Moscow, Russia
ekaterina.kondrateva@skoltech.ru

Abstract. In this work, we aimed at predicting children's fluid intelligence scores based on structural T1-weighted MR images from the largest long term study of brain development and child health. The target variable was regressed on a data collection site, sociodemographic variables, and brain volume, thus being independent to the potentially informative factors, which were not directly related to the brain functioning. We investigated both feature extraction and deep learning approaches as well as different deep CNN architectures and their ensembles. We proposed an advanced architecture of VoxCNNs ensemble, which yields MSE (92.838) on a blind test.

Keywords: MRI analysis · Fluid intelligence prediction · Deep learning · 3D convolutional neural networks · VoxCNN ensemble

1 Introduction

Understanding cognitive development in children may potentially improve their health outcomes through adolescence. Thus, determining neural mechanism underlying general intelligence is a critical task. One of two discrete factors of general intelligence is fluid intelligence.

Fluid intelligence is the capacity to think logically and solve problems in novel situations, independent of acquired knowledge. It involves the ability to identify patterns and relationships that underpin novel problems and to extrapolate these findings using logic [1].

There are studies on fluid intelligence prediction based on various brain imaging techniques and extracted features [23,40]. However, the authors could not highlight robust biomarkers and methods to predict fluid intelligence scores .

Deep learning approaches and convolutional neural networks, in particular, have shown high potential on imagery classification, recognition and processing and thus could be considered useful for fluid intelligence scores prediction based on MRI data (3D brain images).

© Springer Nature Switzerland AG 2019
K. M. Pohl et al. (Eds.): ABCD-NP 2019, LNCS 11791, pp. 158–166, 2019.
https://doi.org/10.1007/978-3-030-31901-4_19

The advantage of deep learning methods is the ability to automatically derive complex and informative features from the raw data during the training process. That allows training a neural network directly on high-dimensional 3D brain imaging data skipping the feature extraction step.

By design, neural architectures for deep learning are built in a modular way, with basic building blocks, such as composite convolutional layers, typically reused across many models and applications. This enables the standardization of deep learning architectures, with much research devoted to the exploration of pre-built layers and pre-trained activations (for transfer learning, image retrieval, etc.). However, the choice of appropriate architecture targeting specific clinical applications such as cognitive potential prediction or pathology classification remains open problem and requires further investigation.

In the present study we carried out an extensive experimental evaluation of deep voxelwise neural network architectures for fluid intelligence scores prediction based on MRI data with multimodal input structure.

The article has the following structure. In Sect. 2 we review deep network architectures used for MRI data processing. In Sect. 3 we present the training dataset and our deep network architecture. We describe obtained results in Sect. 4, provide discussions in Sect. 5 and draw conclusions in Sect. 6.

2   Related Work

There is a number of successful applications of convolutional neural networks (CNN) with different architectures for segmentation of MRI data. Many of these solutions are based on adapting existing approaches to analyzing 2D images for processing of three-dimensional data.

For example, for brain segmentation, an architecture similar to ResNet [20] was proposed, which expands the possibilities of deep residual learning for processing volumetric MRI data using 3D filters in convolutional layers. The model, called VoxResNet [32], consists of volumetric residual blocks (VoxRes blocks), containing convolutional layers as well as several deconvolutional layers. The authors demonstrated the potential of ResNet-like volumetric architectures, achieving better results than many modern methods of MRI image segmentation [22]. Convolutional neural networks also showed good classification results in problems associated with neuropsychiatric diseases such as Alzheimer's disease.

Recently proposed classification model with a VGG-like architecture called VoxCNN was used for neuro-degenerative decease classification [21]. These results were more accurate or comparable to earlier approaches that use previously extracted morphometrical lower dimensional brain characteristics [34,38,39].

Thus, convolutional networks can be applied directly to the raw neuroimaging data without loss of model performance and over-fitting, which allows skipping the pre-processing step.

However, to the depth of our knowledge, there has not been much work on the use of convolutional networks for predicting fluid intelligence based on MRI imaging.

3   Materials and Methods

3.1   Data Set

The training data set was provided by ABCD Neurocognitive Prediction Challenge (ABCD-NP-Challenge 2019[1]). The dataset consists of T1-weighed MR brain images of four thousand individuals (of age 9–10 years) as well as corresponding sociodemographic variables [33]. The participants' fluid intelligence scores (4154 subjects, 3739 for training and 415 for validation) were also provided.

3.2   Target Processing

The fluid intelligence scores were pre-residualized on a data collection site, sociodemographic variables and brain volume. For that a linear regression model was fitted with fluid intelligence as the dependent variable and brain volume, data collection site, age at baseline, sex at birth, race/ethnicity, highest parental education, parental income, and parental marital status as independent variables [33].

The obtained residuals were used as targets to be predicted by a neural network. This approach is known to be used in GML models, for fMRI data analysis, allowing removal of linear dependencies between dependent variables.

3.3   MRI Data Processing

Imagery dataset consists of skull stripped images affinely aligned to the SRI 24 atlas [5], segmented into regions of interest according to the atlas, and the corresponding volume scores of each ROI [29]. T1-weighted MRI was transformed according to the Minimal Processing Pipeline by ABCD [33].

The cross-sectional component of the National Consortium on Alcohol and NeuroDevelopment in Adolescence (NCANDA) pipeline [12] was applied to T1 images. The steps included noise removal and field inhomogeneity correction confined to the brain mask, defined by non-rigidly aligning SRI24 atlas to the T1w MRI via Advanced normalization tools (ANTS) [4].

The brain mask was refined by majority voting across maps extracted by FSL BET [3], AFNI 3dSkullStrip [2], FreeSurfer mrigcut [6], and the Robust Brain Extraction (ROBEX) methods [8], which were applied on combinations of bias and non-bias corrected T1w images. Using the refined masked, image inhomogeneity correction was repeated and the skull-stripped T1w image was segmented into brain tissue (gray matter, white matter, and cerebrospinal fluid) via Atropos [7]. Gray matter tissue was further parcelled according to the SRI24 atlas, which was non-rigidly registered to the T1w image via ANTS.

---

[1] https://sibis.sri.com/abcd-np-challenge/.

3.4   Specifications of the Investigated Models

We use an ensemble of deep neural networks with VoxCNN architecture [27,37] to solve the regression problem. The proposed architecture has already demonstrated some successful applications to brain image analysis tasks. To provide better convergence and stronger regularization of results we enhanced this architecture.

VoxCNN networks are similar to VGG [11] architecture, which is a popular architecture for 2D-images classification. VoxCNN applies 3D convolutions to deal with three-dimensional MRI brain scans.

Proposed network consists of four blocks with two convolutional layers each having 3D convolutions followed by batch-normalization and ReLU activation function [41]. Number of filters in convolutional layers starts from 16 in the first block and doubles with each next block. Filters of the very first layer are applied with the stride x2 to reduce the dimension of the original image. Our experiments have shown that this step does not reduce the network performance but helps to speed up the convergence and meet the limitations of GPU memory. The blocks are separated by max-pooling layers. We also apply 3D-dropout after each pooling layer to promote independence between feature maps and reduce over-fitting [15].

Next, feature maps extracted by the convolutional layers are fed into the fully connected layer with 1024 hidden units, batch-normalization, ReLU activation, and dropout regularization, and then to the final layer with a single unit without non-linearity.

It was previously shown that auxiliary tower backpropagates the classification loss earlier in the network, serving as an additional regularization mechanism [14,24].

Therefore, the auxiliary output was added to the network to provide better training of the deeper layers. For this purpose, feature maps from intermediate layers were fed to the separate fully connected layer to produce another target prediction, which was then added to the main network output with adjusted weight. In this case, the output of the third block of convolutional layer was used to compute auxiliary prediction and average it with the main output with weights 0.4 and 0.6 respectively.

We assessed model quality by Mean Squared Error (MSE) between the predicted scores and the pre-residualized fluid intelligence scores. The models were selected by optimizing the MSE-loss with the Adam optimizer. The learning rate was set to 3e-5, batch size is 10 and each network was trained until the loss on validation set starts to increase.

To train the model we used multi-modal input data: brain scan data (T1-weighted imagery after preprocessing) and gray matter segmented brain masks. For each subject, two three-dimensional images were stacked as channels of a single image. We fed the resulted 3D image with two channels into the VoxCNN network as an input.

We used cross-validation to increase the model performance: we split the training sample into two separate parts and two neural networks are trained with the same architecture on each part independently. Then for the validation subjects, an ensemble of these two models, defined as a weighted average of their predictions, was applied. Weights for averaging were determined based on the validation performance of each model (test predictions of the network that turned out to demonstrate lower MSE score on validation were set to larger weights). The number of layers, Stride and ReLU blocks position were adjusted correspondingly (Fig. 1).

The train set consists of n = 3739 samples, the validation set – n = 415 samples, and the test set – n = 4515 samples.

The models were implemented in PyTorch and trained on a single GPU [18].

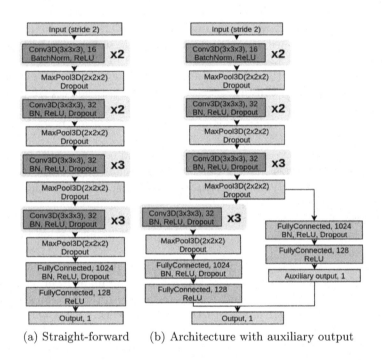

(a) Straight-forward   (b) Architecture with auxiliary output

Fig. 1. VoxCNN model architectures used for fluid target prediction.

4   Experimental Results

In Table 1 the explored deep neural network architectures are specified as well as corresponding results for fluid intelligence prediction. Here the brain morphometric characteristics predictive capacity is considered as a baseline for prediction.

The most accurate prediction (in terms of MSE on the validation set) was obtained as a weighted average of the two predictions by VoxCNN trained on different parts of the training sample:

Table 1. Model architectures and results on the `Validation` set.

#	Model architecture	MSE
1	Brain morphometry	71.293
2	VoxCNN on brain T1 imagery	71.777
3	VoxCNN on 3D segmented brain mask	72.094
4	Ensemble: VoxCNNs on T1 and segmented mask	71.314
5	Ensemble: VoxCNNs on T1, segmented mask with morphology features	70.635

Table 2. Model architectures and results for the fluid intelligence prediction on the `Test` set.

#	Model architecture	MSE
1	Ensemble: VoxCNNs on T1 and segmented mask	92.8378
2	Ensemble: VoxCNNs on T1, segmented mask with morphology features	94.0808

1. VoxCNN network, trained on both brain T1 images and segmented images,
2. VoxCNN network (with auxiliary head for better convergence), trained on brain T1 images, segmented images and additional socio-demographic data. We used segmented brain masks and full brain imagery after pre-processing.

As a result, the first and the second network architectures showed 71.777 and 71.094 `MSE` scores on the `Validation` set. After averaging the predictions with adjusted weights $\frac{2}{3}$ and $\frac{1}{3}$, the final validation performance reached 70.635 MSE when using ensembles of models.

Then on the `Test` set the ensemble models yielded 92.8378 and 94.0808 `MSE` scores correspondingly (Table 2).

5   Discussion

All considered regression models provided `MSE` close to 70. These results are comparable to the baseline result, calculated using morphological characteristics on the `Validation` set.

This incremental improvement and rather high errors across all models could potentially imply both the study design and the data inconsistency: the reason may be that structural T1-weighted images alone are not enough to predict fluid intelligence scores; at the same time brain functional data like fMRI might have more predictive power for cognitive assessment.

The top performing model was a weighted average prediction of two Vox-CNN neural networks trained on different parts of the training sample, highlighting the potential strength of the models' ensembles yielded 70.635 `MSE` on the `Validation` set and 92.635 `MSE` on the `Test` set. Thus combination of different inputs, or so-called data fusion, gives us more information to built accurate prediction. Data fusion models are known to be successful in MRI segmentation applications, for example for epileptical foci detection [26].

6 Conclusion

In our work for the first time ensembles of VoxCNN networks were applied to the 3D brain imagery regression task. According to the results of this architecture we could consider it as a consistent predictive tool for large datasets with heavy and multi-modal inputs.

Due to the complex structure of the considered dataset there is enough room for further improvements. A future work on the model hyperparameters optimization is needed in order to achieve better network convergence. Advanced approaches to initialization of neural network parameters [16] and construction of ensembles [9] could be used. Sparse 3D convolutions could decrease memory requirements [36].

Transfer learning and domain adaptation techniques could potentially show better performance here [19,25,28]. Also it is possible to utilize multi-fidelity approaches when solving the regression problem with multi-modal data [13,30, 31]. Conformal prediction framework [10,17,35] is a ready-to-use tool to assess prediction uncertainty.

Acknowledgements. The work was supported by the Russian Science Foundation under Grant 19-41-04109.

The considered problem was formulated in the scope of the Project "Machine Learning and Pattern Recognition for the development of diagnostic and clinical prognostic prediction tools in psychiatry, borderline mental disorders, and neurology", granted by Skoltech Biomedical Initiative Program, Skolkovo Institute of Science and Technology, Moscow, Russia.

References

1. Carroll, J.B.: Human Cognitive Abilities: A Survey of Factor-Analytic Studies. Cambridge University Press, Cambridge (1993)
2. RobertWCox: AFNI: software for analysis and visualization of functional magnetic resonance neuroimages. In: Computers and Biomedical Research, vol. 29, no. 3, pp. 162–173 (1996)
3. Smith, S.M.: Fast robust automated brain extraction. Hum. Brain Mapp. **17**(3), 143–155 (2002)
4. Avants, B.B., Tustison, N., Song, G.: Advanced normalization tools (ANTS). Insight j **2**, 1–35 (2009)
5. Rohlfing, T., et al.: The SRI24 multichannel atlas of normal adult human brain structure. Hum. Brain Mapp. **31**(5), 798–819 (2010)
6. Sadananthan, S.A., et al.: Skull stripping using graph cuts. NeuroImage **49**(1), 225–239 (2010)
7. Avants, B.B., et al.: An open source multivariate framework for n-tissue segmentation with evaluation on public data. Neuroinformatics **9**(4), 381–400 (2011)
8. Iglesias, J.E., et al.: Robust brain extraction across datasets and comparison with publicly available methods. IEEE Trans. Med. Imaging **30**(9), 1617–1634 (2011)
9. Burnaev, E.V., Prikhod'ko, P.V.: On a method for constructing ensembles of regression models. Autom. Remote Control **74**(10), 1630–1644 (2013)

10. Burnaev, E., Vovk, V.: Efficiency of conformalized ridge regression. In: Balcan, M.F., Feldman, V., Szepesvari, C. (eds.) Proceedings of the 27th Conference on Learning Theory. Proceedings of Machine Learning Research, PMLR, Barcelona, Spain, 13–15 Jun 2014, vol. 35, pp. 605–622 (2014)
11. Simonyan, K., Zisserman, A.: Very deep convolutional networks for large-scale image recognition. arXiv preprint arXiv:1409.1556 (2014)
12. Brown, S.A., et al.: The national consortium on alcohol and neurodevelopment in adolescence (NCANDA): a multisite study of adolescent development and substance use. J. Stud. Alcohol Drugs **76**(6), 895–908 (2015)
13. Burnaev, E., Zaytsev, A.: Surrogate modeling of multifidelity data for large samples. J. Commun. Technol. Electron. **60**(12), 1348–1355 (2015)
14. Szegedy, C., et al.: Going deeper with convolutions. In: Proceedings of the IEEE Conference on Computer Vision and Pattern Recognition, pp. 1–9 (2015)
15. Tompson, J., et al.: Efficient object localization using convolutional networks. In: Proceedings of the IEEE Conference on Computer Vision and Pattern Recognition, pp. 648–656 (2015)
16. Burnaev, E., Erofeev, P.: The influence of parameter initialization on the training time and accuracy of a nonlinear regression model. J. Commun. Technol. Electron. **61**(6), 646–660 (2016). ISSN 1555-6557
17. Burnaev, E., Nazarov, I.: Conformalized Kernel ridge regression. In: 2016 15th IEEE International Conference on Machine Learning and Applications (ICMLA), pp. 45–52 (2016)
18. Canziani, A., Paszke, A., Culurciello, E.: An analysis of deep neural network models for practical applications. arXiv preprint arXiv:1605.07678 (2016)
19. Goetz, M., et al.: DALSA: domain adaptation for supervised learning from sparsely annotated MR images. IEEE Trans. Med. Imaging **35**(1), 184–196 (2016)
20. He, K., et al.: Deep residual learning for image recognition. In: Proceedings of the IEEE Conference on computer vision and pattern recognition, pp. 770–778 (2016)
21. Hosseini-Asl, E., Gimel'farb, G., El-Baz, A.: Alzheimer's disease diagnostics by a deeply supervised adaptable 3D convolutional network. arXiv preprint arXiv:1607.00556 (2016)
22. Milletari, F., Navab, N., Ahmadi, S.-A.: V-net: fully convolutional neural networks for volumetric medical image segmentation. In: 2016 Fourth International Conference on 3D Vision (3DV), pp. 565–571. IEEE (2016)
23. Paul, E.J., et al.: Dissociable brain biomarkers of UID intelligence. NeuroImage **137**, 201–211 (2016)
24. Szegedy, C., et al.: Rethinking the inception architecture for computer vision. In: Proceedings of the IEEE Conference on Computer Vision and Pattern Recognition, pp. 2818–2826 (2016)
25. Ghafoorian, M., et al.: Transfer learning for domain adaptation in MRI: application in brain lesion segmentation. In: Descoteaux, M., Maier-Hein, L., Franz, A., Jannin, P., Collins, D.L., Duchesne, S. (eds.) MICCAI 2017. LNCS, vol. 10435, pp. 516–524. Springer, Cham (2017). https://doi.org/10.1007/978-3-319-66179-7_59
26. Hunyadi, B., et al.: Tensor decompositions and data fusion in epileptic electroencephalography and functional magnetic resonance imaging data. Wiley Interdiscip. Rev.: Data Min. Knowl. Discov. **7**(1), e1197 (2017)
27. Korolev, S., et al.: Residual and plain convolutional neural networks for 3D brain MRI classification. In: IEEE 14th International Symposium on Biomedical Imaging (ISBI 2017), pp. 835–838. IEEE (2017)

28. Lu, H., et al.: When unsupervised domain adaptation meets tensor representations. In: Proceedings of the IEEE International Conference on Computer Vision, pp. 599–608 (2017)
29. Pfefferbaum, A., et al.: Altered brain developmental trajectories in adolescents after initiating drinking. Am. J. Psychiatry **175**(4), 370–380 (2017)
30. Zaytsev, A., Burnaev, E.: Large scale variable fidelity surrogate modeling. Ann. Math. Artif. Intell. **81**(1), 167–186 (2017). ISSN 1573-7470
31. Zaytsev, A., Burnaev, E.: Minimax approach to variable fidelity data interpolation. In: Singh, A., Zhu, J. (eds.) Proceedings of the 20th International Conference on Artificial Intelligence and Statistics, Proceedings of Machine Learning Research, PMLR, Fort Lauderdale, FL, USA, 20–22 Apr 2017, vol. 54, pp. 652–661 (2017)
32. Chen, H., et al.: VoxResNet: deep voxelwise residual networks for brain segmentation from 3D MR images. NeuroImage **170**, 446–455 (2018)
33. Hagler, D.J., et al.: Image processing and analysis methods for the adolescent brain cognitive development study. bioRxiv, p. 457739 (2018)
34. Ivanov, S., et al.: Learning connectivity patterns via graph kernels for fMRI-based Depression Diagnostics. In: Proceedings of IEEE International Conference on Data Mining Workshops (ICDMW), pp. 308–314 (2018)
35. Kuleshov, A., Bernstein, A., Burnaev, E.: Conformal prediction in manifold learning. In: Gammerman, A., et al. (eds.) Proceedings of the Seventh Workshop on Conformal and Probabilistic Prediction and Applications, Proceedings of Machine Learning Research, PMLR, vol. 91. pp. 234–253 (2018)
36. Notchenko, A., Kapushev, Y., Burnaev, E.: Large-scale shape retrieval with sparse 3D convolutional neural networks. In: van der Aalst, W.M.P., et al. (eds.) AIST 2017. LNCS, vol. 10716, pp. 245–254. Springer, Cham (2018). https://doi.org/10.1007/978-3-319-73013-4_23
37. Pominova, M., et al.: Voxelwise 3D convolutional and recurrent neural networks for epilepsy and depression diagnostics from structural and functional MRI Data. In: 2018 IEEE International Conference on Data Mining Workshops (ICDMW), pp. 299–307. IEEE (2018)
38. Sharaev, M., et al.: MRI-based diagnostics of depression concomitant with epilepsy: in search of the potential biomarkers. In: Proceedings of IEEE 5th International Conference on Data Science and Advanced Analytics, pp. 555–564 (2018)
39. Sharaev, M., et al.: Pattern recognition pipeline for neuroimaging data. In: Pancioni, L., Schwenker, F., Trentin, E. (eds.) ANNPR 2018. LNCS (LNAI), vol. 11081, pp. 306–319. Springer, Cham (2018). https://doi.org/10.1007/978-3-319-99978-4_24
40. Zhu, M., Liu, B., Li, J.: Prediction of general fluid intelligence using cortical measurements and underlying genetic mechanisms. In: IOP Conference Series: Materials Science and Engineering, vol. 381, no. 1, p. 012186. IOP Publishing (2018)
41. Eckle, K., Schmidt-Hieber, J.: A comparison of deep networks with ReLU activation function and linear spline-type methods. Neural Netw. **110**, 232–242 (2019)

Adolescent Fluid Intelligence Prediction from Regional Brain Volumes and Cortical Curvatures Using BlockPC-XGBoost

Tengfei Li[1], Xifeng Wang[2], Tianyou Luo[2], Yue Yang[2], Bingxin Zhao[2], Liuqing Yang[3], Ziliang Zhu[2], and Hongtu Zhu[2(✉)]

[1] Department of Radiology and the Biomedical Research Imaging Center, The University of North Carolina at Chapel Hill, Chapel Hill, NC, USA
[2] Department of Biostatistics, The University of North Carolina at Chapel Hill, Chapel Hill, NC, USA
htzhu@email.unc.edu
[3] Department of Statistics and Operations Research, The University of North Carolina at Chapel Hill, Chapel Hill, NC, USA

Abstract. From the ABCD dataset, we discover that besides the gray matter volume of cortical regions, other measures such as the mean cortical curvature, white matter volume and subcortical volume exhibit additional capabilities in the prediction of the pre-residulized fluid intelligence scores for adolescents. The MSE and R-square on validation dataset are improved from 70.65 and 0.0175 to 69.39 and 0.0350, respectively, comparing with using mostly the grey matter volume provided by the challenge organizer. Specifically, by employing a BlockPC-XGBoost framework we discover the following predictors in reducing the MSE on validation set: the gray matter volume of right posterior cingulate gyrus and left caudate nucleus, the entorhinal white matter volume of the left hemisphere, the number of detected surface holes, the globus pallidus volume, the mean curvatures of precentral gyrus, postcentral gyrus and Banks of Superior Temporal Sulcus.

Keywords: Fluid intelligence · ABCD · Block PCA · XGBoost

1 Introduction

Fluid intelligence is one crucial component of human general intelligence which involves the capacity to think logically, solve problems in novel situations and independent of acquired knowledge [14]. It has been widely accepted that the fluid intelligence reaches a peak in late adolescence and then declines [13]. Furthermore, its quantification and accurate predictions are important for teenagers which foresees creative achievement, scholastic performance, employment prospects, socioeconomic status, etc. in their future years [8]. Structural magnetic

© Springer Nature Switzerland AG 2019
K. M. Pohl et al. (Eds.): ABCD-NP 2019, LNCS 11791, pp. 167–175, 2019.
https://doi.org/10.1007/978-3-030-31901-4_20

resonance (MR) images are one of the most powerful tool to help predict the fluid intelligence. The ABCD challenge dataset provides us with a large amount of adolescent participants with structural MR images, aiming at the precise prediction of the pre-Residualized Fluid Intelligence Scores (RFIS), which has been adjusted for different data collection sites, demographic variables and whole brain volume. It includes 3,739 training subjects, 415 validation subjects and 4402 test subjects. For each subject, the cortical volume of 122 regions of interest were extracted from T1 images by the challenge organizers according to the NCANDA pipeline [20].

From previous literature, scientists found that regional brain volumes were closely related to the cognitive status of individuals from the study of Alzheimer disease [2]. While most of the previous literature mainly focus on gray matter cortical metrics [3,23], recent studies revealed the association of white matter and subcortical volumes with cognitive functions [11,25]. Cortical thickness can also be related to the cognition according to [7,15]. Based on those findings we extracted white matter, subcortical regional volumes, etc., by pre-processing the raw T1 MR images using the software Freesurfer [21] to obtain more resourceful explanatory features in different regions of interest (ROIs) for improvement of the prediction performance. We discovered that besides the cortical volumes, structural metrics such as white matter and subcortical volume and mean curvatures were also useful based on our challenge results.

The information extraction and manipulation are two important components for prediction. For information extraction, principal component analysis (PCA) is a common and standard technique in multivariate statistics, which aims to use a set of linear approximations for dimension reduction. However, the classical procedure of PCA could lead to a side effect that the principal components mainly focus on subdomains with large variance. When small groups of highly-correlated covariates exist, important features might get hidden behind [16]. Human brain encompasses complex network structures, and different brain regions can be highly-correlated. Cluster analysis is a typical method of grouping a set of objects into different subsets in terms of their "similarities". Previous literature has shown that hierarchical clustering results could correspond to brain anatomical configurations [17]. By dividing similar covariates into groups to extract principle components (PCs) within each cluster, we can preserve important features. For information manipulation, statistical modeling and machine learning (ML) methods, such as linear and ridge regression [12], random forest [5], support vector machines [9], etc., remain popular for decades. Among those, the Extreme Gradient Boosting (XGBoost) [6] is extensively used by ML practitioners to create state of art data science solutions, and has gained much attention recently as the choice of many winning teams of ML competitions [24].

We hereby propose to use hierarchical clustering with block PCA to extract important features which are fed into the XGBoost machine for predicting fluid intelligence. Our results show that incorporating block PCA into XGBoost framework leads to better prediction performance than using XGBoost based on either original covariates or traditionally calculated principal components.

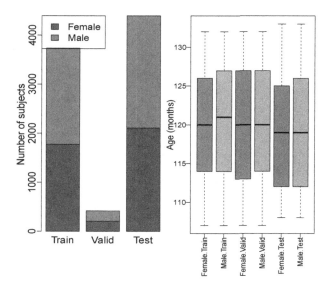

Fig. 1. Age and gender distribution (left) and number of subjects among training, validation and test datasets.

2 Method

2.1 Dataset

The ABCD challenge uses NIH Toolbox® [19], Rey Auditory Verbal Learning Test (RAVLT) [10], Little man task [1], etc., to quantify the fluid intelligence scores [22]. The whole dataset we obtained includes 4459 males, 4085 females aged from 8 to 12 years old (107–133 months), and 12 additional individuals with missing demographic information. A detailed distribution of ages and genders for each of the training, validation and test set can be found in Fig. 1. Raw T1-weighted MR images were multi-protocol acquired with Siemans, Philips and GE scanners, which were further processed according to [20]. The cortical ROI volumes for each subject were calculated. With this dataset we describe our workflow in the following subsections and illustrated it in Fig. 4(a).

2.2 The Preprocessing

To predict the RFIS of subjects on validation and test set, we first pre-processed the raw T1 MR images for participants in the whole dataset to extract their brain white matter and subcortical ROI volumes, and the mean curvatures, etc., by using the software FREESURFER's standard recon-all pipeline (v.6.0.0) [24], which include motion correction, intensity normalization, skull stripping, removal of non-brain tissue, brain mask generation, cortical reconstruction, WM and subcortical segmentation, and cortical parcellation. The white matter volume, subcortical volume, and mean curvature for each ROI and individual were

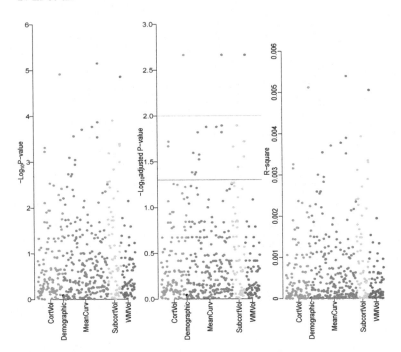

Fig. 2. Manhattan plots: $-\log_{10}$ P-value (left), $-\log_{10}$ FDR adjusted P-value (middle) and the square R^2 (right) of Pearson's correlation between cortical volumes, two demographic covariates (age and gender), mean curvatures and subcortical volumes and the response RFIS.

extracted as supplemental information which were combined with cortical volumes, age and gender in Subsect. 2.1 to make predictions. Pearson's correlations and p-values between all structural metrics with the RFIS were calculated from all subjects in training dataset and displayed in Fig. 2.

We discovered that the age and gender were not significant (p-value > 0.80); there were 22 features with the False Discovery Rate (FDR) adjusted p-values [4] smaller than 0.05, which include: white matter volume of pons and left entorhinal; gray matter volume of left and right parahippocampal gyrus; subcortical volume of right and left globus pallidus and right ventral diencephalon; mean cortical curvature of right and left precentral gyrus, right postcentral gyrus, right paracentral lobule, right and left superior parietal lobule, right Banks of Superior Temporal Sulcus, right superior temporal gyrus, right medial orbital gyrus and right inferior temporal gyrus; number of defect holes in right and total surface. The top 3 features were white matter volume of pons gray matter of right precentral gyrus and white matter volume of left entorhinal, whose FDR adjusted p-values were less than 0.01.

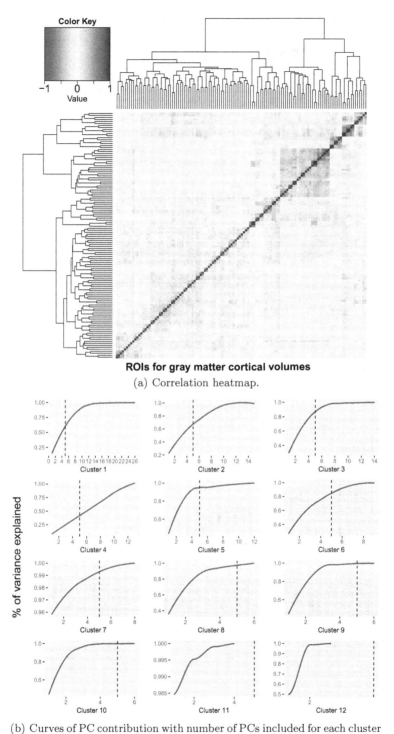

(a) Correlation heatmap.

(b) Curves of PC contribution with number of PCs included for each cluster

Fig. 3. Clustering analysis and PCA based on all the regional cortical volumes

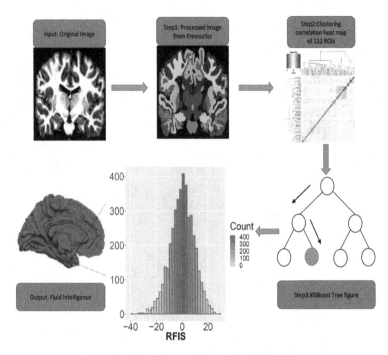

(a) Workflows of the BlockPC-XGB.

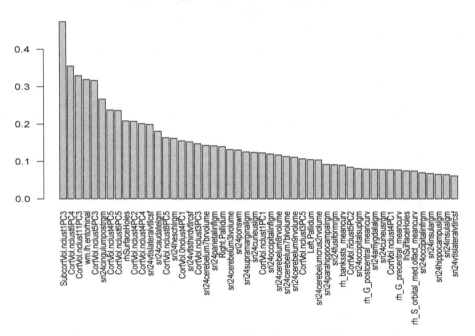

(b) The importance scores of the top 50 features for XGBoost prediction.

Fig. 4. Workflow and the important predictors.

2.3   Clustering and Block PCA

To efficiently extract information from the original datasets, We first perform hierarchical clustering using WardD2 algorithm [18] on the four structural metrics: cortical volumes, WM volumes, subcortical volumes and mean curvature, separately. Looking at the clustering results, we found that cutting the number of clusters at 12 led to the best performance. For each cluster we extracted the first 5 PCs. Hence there could be totally about $12 \times 5 \times 4 = 240$ additional features extracted if 5 PCs for each cluster exists. However, in real data there were only 206 PCs generated due to the fact that for very small clusters, the total number of principal components is less than 5. Those additional features were then combined with the original ROI quantitative measures in previous steps to fit the XGBoost model for prediction. In Fig. 3 we provided one example to illustrate the correlation heatmap of the clustering structures for cortical volumes, which indicates a latent correlation structure among all the regions. The Figure also shows that the first 5 principle components for almost all clusters cover more than 60% contribution of variance. Combining the 206 PCs with the original features generated from the previous steps, Pearson's correlations and p-values with the RFIS response were calculated similar to Subsect. 2.2. After the FDR p-value correction, 30 features were found significantly correlated with the RFIS, with adjusted p-values less than 0.05.

2.4   XGBoost Statistical Models

All features generated in the last three steps on the training set were combined together as explanary varibles to fit a prediction model. First, pvalues based on Pearson's correlation between all features and the RFIS by only using the training dataset were ranked, where the first p_0 features with the smallest p-values were screened which was then fed into the XGBoost machine. The "GbLinear" booster was used with the default "reg:linear" objective function and the initial prediction score ("base_score") was set as zero; we set the learning rate as 0.05, and used 10-fold cross validation on the training dataset with the mean absolute error (MAE) as the evaluation metric to select the number of iteration rounds. The L_2 tuning parameter was fixed at the default value 1. We trained XGBoost models with different tuning parameters and then made predictions on the validation set. The optimal p_0 was then selected to minimize the mean squared error (MSE) on the validation set.

We made comparisons between the BlockPC-XGBoost with the XGBoost without Block PCs, and between the BlockPC-XGBoost using all features with BlockPC-XGBoost using the cortical volumes provided by the challenge organizer. The results for both training and validation were shown in the next section.

3   Results

The MSE and R square for the BlockPC-XGBoost using all features, the BlockPC-XGBoost using mostly the grey matter cortical volume provided by the challenge organizer, and the XGBoost without Block PCs for both training and validation were shown in Table 1. The comparisons indicate the advantage of the proposed BlockPC-XGBoost. Importance scores of the top features for the BlockPC-XGBoost were shown in Fig. 4(b). We found that the important predictors did not match the significant features in Fig. 2 exactly. These important predictors were: the gray matter volume of right posterior cingulate gyrus and left caudate nucleus, the entorhinal white matter volume of the left hemisphere, the number of detected surface holes, the globus pallidus volume, several regional lateral ventricle and cerebellum volumes, the mean curvatures of precentral gyrus, postcentral gyrus and Banks of Superior Temporal Sulcus. Among those, the 15 PCs from clustering takes 30% of the top 50 features in total.

Table 1. Prediction results using 3 methods

Method	Train: MSE	Train: R-square	Val: MSE	Val: R-square	Test: MSE
BlockPC-XGB (All)	82.508	0.0412	69.386	0.0350	93.156
BlockPC-XGB (Cort)	83.236	0.0321	70.652	0.0175	–
XGB (All)	82.107	0.0486	69.971	0.0264	–

4   Discussion

From the analysis discussed above, we found several brain areas (white matter volume of pons, gray matter of right precentral gyrus and white matter volume of left entorhinal, etc.) significantly correlated with RFIS. Based on the given features from the challenge organizer and our generated features by Freesurfer software, we used a Block PCA design to learn the representation from all these features, which shows a good learning ability for correlated features. We then used the XGBoost machine to train a prediction model using the learned features, obtaining a result of 69.39 on the validation set. Simultaneously, we found several features which exhibit strong prediction power.

However, the proposed approach is based on segmentation and parcellation of the ROIs, which relies on the image processing precision. Furthermore, the approach does not consider spatial location of all ROIs. Comparing with modern deep learning techniques, e.g., the U-Net, or the graphical model-based deep neural network, it loses local information.

References

1. Acker, W., Acker, C.: Bexley maudsley automated processing screening and bexley maudsley category sorting test manual. NFER-Nelson, Windsor, England (1982)
2. Adak, S., et al.: Predicting the rate of cognitive decline in aging and early alzheimer disease. Neurology **63**(1), 108–114 (2004)
3. Bajaj, S., et al.: The relationship between general intelligence and cortical structure in healthy individuals. Neuroscience **388**, 36–44 (2018)

4. Benjamini, Y., Hochberg, Y.: Controlling the false discovery rate: a practical and powerful approach to multiple testing. J. R. Stat. Soc. Ser. B **57**, 289–300 (1995)
5. Breiman, L.: Random forests. Mach. Learn. **45**(1), 5–32 (2001)
6. Chen, T., Guestrin, C.: Xgboost: a scalable tree boosting system. In: Proceedings of the 22nd ACM SIGKDD International Conference on Knowledge Discovery and Data Mining, pp. 785–794 (2008)
7. Cheng, C., Cheng, S., Tam, C., Chan, W., Chu, W., Lam, L.: Relationship between cortical thickness and neuropsychological performance in normal older adults and those with mild cognitive impairment. Aging Dis. **9**(6), 1020–1030 (2018)
8. Colom, R., Escorial, S., Shih, P., Privado, J.: Fluid intelligence, memory span, and temperament difficulties predict academic performance of young adolescents. Pers. Individ. Differ. **42**(8), 1503–1514 (2007)
9. Cortes, C., Vapnik, V.: Support-vector networks. Mach. Learn. **20**(3), 273–297 (1995)
10. Daniel, M., Wahlstrom, D., Zhang, O.: Equivalence of q-interactive^{TM} and paper administrations of cognitive tasks: Wisc-v. Q-Interactive Technical Report 8 (2014)
11. Davies, G., et al.: Study of 300,486 individuals identifies 148 independent genetic loci influencing general cognitive function. Nat. Commun. **9**(1), 2098 (2018)
12. Hoerl, A.E., Kennard, R.W.: Ridge regression: biased estimation for nonorthogonal problems. Technometrics **12**(1), 55–67 (1970)
13. Horn, J.L.: Human Ability Systems. Academic Press, New York (1978)
14. Jaeggi, S.M., Buschkuehl, M., Jonides, J., Perrig, W.J.: Improving fluid intelligence with training on working memory. In: Proceedings of the National Academy of Sciences of the United States of America, vol. 105, pp. 6829–6833 (2008)
15. Jiang, L., et al.: Cortical thickness changes correlate with cognition changes after cognitive training: evidence from a Chinese community study. Front. Aging Neurosci. **8**, 118–118 (2016)
16. Lin, Z., Zhu, H.: MFPCA: multiscale functional principal component analysis. In: The Thirty-Third AAAI Conference on Artificial Intelligence (AAAI 2019) (2019)
17. Menon, D., Bullmore, E., Suckling, J., Pickard, J.D., Coleman, M.R., Salvador, R.: Neurophysiological architecture of functional magnetic resonance images of human brain. Cereb. Cortex **15**(9), 1332–1342 (2005)
18. Murtagh, F., Legendre, P.: Ward's hierarchical agglomerative clustering method: which algorithms implement ward's criterion? J. Classif. **31**(3), 274–295 (2014)
19. NIH Toolbox. http://www.nihtoolbox.org
20. Pfefferbaum, A., et al.: Altered brain developmental trajectories in adolescents after initiating drinking. Am. J. Psychiatry **175**(4), 370–380 (2017)
21. Reuter, M., Schmansky, N.J., Rosas, H.D., Fischl, B.: Within-subject template estimation for unbiased longitudinal image analysis. NeuroImage **61**(4), 1402–1418 (2012)
22. Thompson, W., et al.: The structure of cognition in 9 and 10 year-old children and associations with problem behaviors: findings from the ABCD study's baseline neurocognitive battery. Dev. Cogn. Neurosci. **36**, 100606 (2018)
23. Walhovd, K., et al.: Cortical volume and speed-of-processing are complementary in prediction of performance intelligence. Neuropsychologia **43**(5), 704–713 (2005)
24. XGBoost - ML winning solutions. https://github.com/dmlc/xgboost/tree/master/demo#machine-learning-challenge-winning-solutions
25. Zhao, B., et al.: Gwas of 19,629 individuals identifies novel genetic variants for regional brain volumes and refines their genetic co-architecture with cognitive and mental health traits. bioRxiv (2019)

Cortical and Subcortical Contributions to Predicting Intelligence Using 3D ConvNets

Yukai Zou[1,2(✉)], Ikbeom Jang[3], Timothy G. Reese[4], Jinxia Yao[1],
Wenbin Zhu[4], and Joseph V. Rispoli[1,3,5]

[1] Weldon School of Biomedical Engineering, Purdue University, West Lafayette, USA
`{zou75,yao150,jrispoli}@purdue.edu`
[2] College of Veterinary Medicine, Purdue University, West Lafayette, USA
[3] School of Electrical and Computer Engineering, Purdue University,
West Lafayette, USA
`jang69@purdue.edu`
[4] Department of Statistics, Purdue University, West Lafayette, USA
`{reese18,zhu633}@purdue.edu`
[5] Institute of Inflammation, Immunology and Infectious Disease, Purdue University,
West Lafayette, USA

Abstract. We present a novel framework using 3D convolutional neural networks to predict residualized fluid intelligence scores in the MICCAI 2019 Adolescent Brain Cognitive Development Neurocognitive Prediction Challenge datasets. Using gray matter segmentations from T1-weighted MRI volumes as inputs, our framework identified several cortical and subcortical brain regions where the predicted errors were lower than random guessing in the validation set (mean squared error = 71.5252), and our final outcomes (mean squared error = 70.5787 in the validation set, 92.7407 in the test set) were comprised of the median scores predicted from these regions.

Keywords: Adolescence · Brain · Convolutional neural networks · Fluid intelligence · MRI

1 Introduction

Unraveling puzzles between behavior and human brain has long been an intriguing topic in cognitive neuroscience [16]. One important research question is to understand how intelligence relates to brain structure in adolescence. There is evidence showing that fluid intelligence [2], the capacity of learning and adapting to novel situations, improves rapidly during late childhood (age 8–15) and is thought to be primarily influenced by neurobiological factors [1,12]. Derived from a collection of gold standard tests, the fluid intelligence scores are continuous values with normal distribution, thus posing a very interesting challenge

© Springer Nature Switzerland AG 2019
K. M. Pohl et al. (Eds.): ABCD-NP 2019, LNCS 11791, pp. 176–185, 2019.
https://doi.org/10.1007/978-3-030-31901-4_21

in machine learning: can fluid intelligence be predicted from high dimensional features, such as brain morphometry?

Carrying this idea to a broader scientific community, a group of researchers at Stanford University, UC San Diego, Vanderbilt University, and Children's National Health System initiated the MICCAI 2019 ABCD Neurocognitive Prediction Challenge. Participating teams of the Challenge proposed algorithms that predict the residualized fluid intelligence scores using T1-weighted (T1w) MRI. The original fluid intelligence scores were residualized to remove confounding factors (brain volume, data collection site, and socio-demographic variables); the residualized scores and T1w brain images of 4,154 participants (3,739 for training set, 415 for validation set) were provided, whereas the scores of 4,515 participants (test set) were predicted from their images.

In neuroimaging, several machine learning methods have been proposed for predicting single continuous values from high feature dimensionality. Wang et al. [20] proposed a sparse learning framework using a Support Vector Regression model to predict the Intelligence Quotient from structural MRI; the framework reduced dimensionality by selecting the derived gray and white matter features, and by fusing different features, the multi-kernel model achieved better performance than the single-kernel model. In 2017, Cole et al. [3] first demonstrated that by applying 3D convolutional neural networks (ConvNets), the chronological age of healthy individuals can be reliably predicted based on T1w brain images that were only minimally processed, suggesting that 3D ConvNets have strengths in high-dimensional prediction tasks and discovering potential relationships between neuroimages and behavioral outcomes [14].

Inspired by the previous work, here we propose a novel framework for the intelligence prediction task on ABCD datasets, using 3D ConvNets trained on multiple cortical and subcortical brain regions. We hypothesize that under our framework, (1) certain brain regions can predict residualized fluid intelligence scores (a.k.a. predictive regions), and (2) compared to each predictive region, the median predicted scores from multiple predictive regions contribute to a lower mean squared error (MSE). Below we describe the methods and preliminary results in detail.

2   Materials and Methods

2.1   Dataset

The ABCD Study [10] is by far the largest multisite longitudinal study of brain development and child health in the United States. The Study has recruited over 11,500 children ages 9–10 at 21 sites across the country. In February 2018, the first annual curated ABCD data (Release 1.0) were made available on NIMH Data Archive (NDA), including minimally processed brain image volumes and tabulated results of structural MRI, diffusion MRI, and fMRI (both resting-state and task-based). Non-imaging assessments were also provided, including physical and mental health, neurocognition, substance use, biospecimens, as well as culture and environment domains.

The ABCD data used in this report came from Release 1.1. For the training set (3,739 participants) and validation set (415 participants), residualized fluid intelligence scores and processed T1w MRI were provided; for the test set (4,515 participants), only the images were provided. The fluid intelligence scores were collected in NIH Toolbox Neurocognition Battery [1]. The raw T1w brain images were acquired using a 3D T1w inversion prepared RF-spoiled gradient echo scan (1 mm isotropic), with prospective motion correction [19,21].

2.2   Processing

Fluid Intelligence Scores. The fluid intelligence scores were residualized using a linear regression model, with brain volume, data collection site, age at baseline, sex at birth, race/ethnicity, highest parental education, parental income, and parental marital status as independent variables. Any participant in the training or validation set with missing values in the dependent or independent variables was excluded. After model fitting, the residuals were computed for all the participants. The R code implementing the procedure has been made available on the official website (https://sibis.sri.com/abcd-np-challenge/).

T1-Weighted (T1w) MRI. A detailed documentation of MRI processing can be found in Pfefferbaum et al.'s Data Supplement [13]. First, the raw data were transformed into NIfTI formats [8], followed by noise removal, field inhomogeneity correction, and confined to a brain mask defined by non-rigidly aligning SRI24 atlas [15] to the T1w MRI. The brain mask was refined by a majority voting approach among the outputs of a variety of neuroimaging software. Based on the refined masks, inhomogeneity correction was repeated, and the skull-stripped T1w image was segmented into gray matter, white matter, and cerebrospinal fluid. Based on the SRI24/TZO parcellation map, the gray matter tissue was further parcellated after non-rigidly aligning the T1w image to the SRI24 atlas. Afterwards, skull-stripped T1w image and corresponding gray matter segmentations were affinely mapped to the SRI24 atlas, and the results were visually inspected.

2.3   3D ConvNets Framework

A schematic illustration of the 3D ConvNets framework is shown in Fig. 1. Due to memory constraints, we proposed this framework after the attempts of feeding the whole brain into the 3D ConvNet and training volumes of multiple brain regions in a simultaneous manner. First, the whole brain volume of size ($240 \times 240 \times 240$) was trimmed down to the specific size and location of each region of interest (ROI) according to the gray matter segmentations, as summarized in Table 1. The gray matter within each ROI was taken as input of 3D ConvNet. The ConvNet contains 3 repeated blocks of: a ($3 \times 3 \times 3$) convolutional layer (with stride of 1 and an L2 regularizer of 1×10^{-4}), a 3D batch-normalization layer [9], an ELU activation function, a ($2 \times 2 \times 2$) average pooling layer (with stride

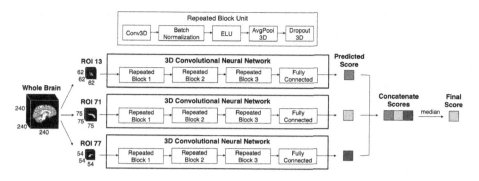

Fig. 1. Schematic illustration of the proposed 3D ConvNets framework. Only the three most predictive regions that contributed to the final submission are shown. ROI 13: left triangular part of interior frontal gyrus; ROI 71: left caudate; ROI 77: left thalamus.

of 2), and a 3D dropout layer (rate of 0.4 for the first two blocks, 0.3 for the third) [18]. The number of feature channels for the three blocks were 8, 16, and 32 respectively. The total number of parameters for the three blocks were 224, 3,472, and 13,856 respectively. For the three ROIs (left triangular part of interior frontal gyrus, left caudate nucleus, and left thalamus) that were included for the final prediction scores, the number of parameters in the last fully connected layer were 6,913, 10,977, and 4,001 respectively. The predicted scores from these ROIs were concatenated to obtain the median predicted scores, which contributed to the final scores.

2.4   Selected Brain Regions

We focused on the brain areas previously reported by Wang et al. [20] for estimating intelligence, including: bilateral transverse temporal gyri, bilateral thalamus (left is shown in Fig. 2), left parahippocampal gyrus, left hippocampus, right opercular part of inferior frontal gyrus, left anterior cingulate gyrus, right amygdala, left lingual gyrus, left superior parietal lobule, right inferior parietal lobule, left angular gyrus, left paracentral lobule, and left caudate nucleus (shown in Fig. 2). For those areas that were unilateral, we also explored their contralateral part. In addition, we explored bilateral triangular part of inferior frontal gyri (left is shown in Fig. 2). The full list of regions can be found in Table 1.

2.5   Implementation

During training, we formed batches by randomly sampling from the 3D volumes, with a batch size of 32. 3 epochs were chosen for training all the selected ROIs. The weights were trained by minimizing the MSE using the Adam optimizer [11], with a learning rate of 0.1 and constant decay of 5×10^{-5} after each epoch.

To determine ROIs that predict the scores in the validation set well, all the MSEs were compared to a "random guessing" model, which is essentially an

Table 1. Specific dimensions and locations of the selected brain regions in the SRI24 space [15]. L/R: left/right hemisphere.

ROI	Brain Region	Dimensions	X		Y		Z	
			Min	Max	Min	Max	Min	Max
11	L Inferior frontal gyrus - opercular	$65 \times 65 \times 65$	41	101	105	165	102	162
12	R Inferior frontal gyrus - opercular	$65 \times 65 \times 65$	131	196	95	160	96	161
13	L Inferior frontal gyrus - triangular	$62 \times 62 \times 62$	40	102	127	189	100	162
14	R Inferior frontal gyrus - triangular	$63 \times 63 \times 63$	130	193	126	189	95	158
31	L Cingulate gyrus - anterior	$83 \times 83 \times 83$	51	134	113	196	89	172
32	R Cingulate gyrus - anterior	$80 \times 80 \times 80$	71	151	115	195	96	176
37	L Hippocampus	$64 \times 64 \times 64$	57	121	76	140	64	128
38	R Hippocampus	$64 \times 64 \times 64$	107	171	76	140	66	130
39	L Parahippocampal gyrus	$83 \times 83 \times 83$	40	123	74	157	35	118
40	R Parahippocampal gyrus	$78 \times 78 \times 78$	89	167	75	153	39	117
41	L Amygdala	$41 \times 41 \times 41$	78	119	105	146	76	117
42	R Amygdala	$44 \times 44 \times 44$	120	164	105	149	75	119
47	L Lingual gyrus	$76 \times 76 \times 76$	55	131	25	101	48	124
48	R Lingual gyrus	$78 \times 78 \times 78$	90	168	24	102	47	125
59	L Parietal lobule - superior	$75 \times 75 \times 75$	48	123	16	91	121	196
60	R Parietal lobule - superior	$75 \times 75 \times 75$	91	166	15	90	121	196
61	L Parietal lobule - inferior	$56 \times 56 \times 56$	49	105	47	103	122	178
62	R Parietal lobule - inferior	$74 \times 74 \times 74$	124	198	32	106	106	180
65	L Angular gyrus	$68 \times 68 \times 68$	40	108	19	87	104	172
66	R Angular gyrus	$63 \times 63 \times 63$	134	197	15	78	105	168
69	L Paracentral lobule	$65 \times 65 \times 65$	68	133	43	108	135	200
70	R Paracentral lobule	$72 \times 72 \times 72$	81	153	38	110	129	201
71	L Caudate nucleus	$75 \times 75 \times 75$	56	131	89	164	77	152
72	R Caudate nucleus	$70 \times 70 \times 70$	84	154	93	163	83	153
77	L Thalamus	$54 \times 54 \times 54$	74	128	81	135	88	142
78	R Thalamus	$54 \times 54 \times 54$	98	152	82	136	89	143
79	L Temporal gyrus - transverse	$56 \times 56 \times 56$	45	101	80	136	84	140
80	R Temporal gyrus - transverse	$65 \times 65 \times 65$	132	197	62	127	77	142

MSE computed after assigning the mean of the residualized fluid intelligence score to each individual, and only those ROIs whose MSE was lower than the random guessing model were selected to compute the median predicted scores (final scores).

The 3D ConvNets were implemented using Keras library (v2.2.4) with Tensorflow (v1.12) as the backend in Python (v3.6) environment. The cluster consisted of 16 Intel Xeon processors, 196 GB system RAM, and two NVIDIA Tesla

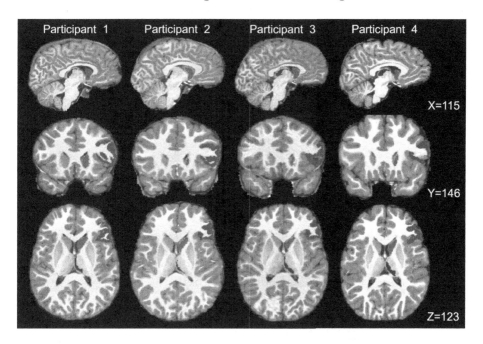

Fig. 2. Anatomical illustrations of the three most predictive ROIs among four participants. Red: left caudate nucleus (ROI 71); Green: left triangular part of interior frontal gyrus (ROI 13); Blue: left thalamus (ROI 77). (Color figure online)

P100 GPUs (each with 16 GB memory). The GPUs were used independently to optimize the ConvNets; once the optimization was done, one GPU was used to evaluate the test set.

3 Preliminary Results

Different number of epochs were tested, and 3 epochs were chosen for training the 28 selected brain regions (see Fig. 3). The total time used for training the 28 3D ConvNets was approximately 6 h.

For the validation set, the mean and standard deviation for the residualized fluid intelligence scores was -0.50 ± 8.47. The MSE of the random guessing model was 71.5252. Among all the prediction MSEs in validation, the three regions producing lower prediction error than the random guessing model were: left caudate nucleus (ROI 71) (MSE = 70.9454, $R^2 = 0.0451$), left triangular part of inferior frontal gyrus (ROI 13) (MSE = 71.1361, $R^2 = 0.0060$), and left thalamus (ROI 77) (MSE = 71.2036, $R^2 = 0.0068$). The median predicted scores resulted from these three ROIs produced a prediction error of 70.5787 ($R^2 = 0.0323$).

In addition, we observed that the right amygdala (ROI 42) produced a slightly lower prediction error (71.4737, $R^2 = 0.0029$) than the random guessing model, and we explored including it in our model; since the final prediction error did

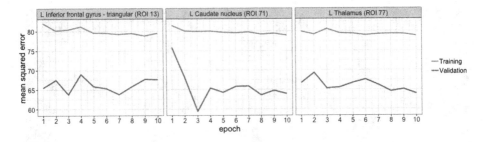

Fig. 3. Prediction errors during training (batch size=32). Results of ten epochs are shown. Left: left triangular part of inferior frontal gyrus (ROI 13); Center: left caudate nucleus (ROI 71); Right: left thalamus (ROI 77).

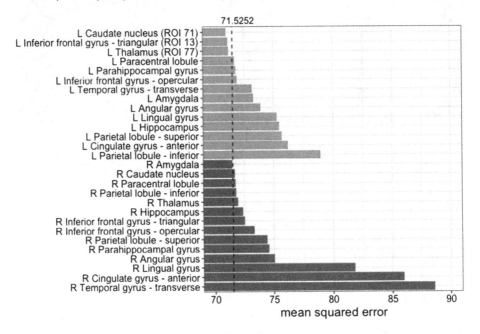

Fig. 4. Prediction errors of all the selected brain regions in the validation set, in comparison with the random guessing model (dashed line).

not improve (i.e., lower than 70.5787), we decided to exclude this ROI from our final model. The prediction errors in the validation set are summarized in Fig. 4. In the end, our proposed framework produced a prediction error of 92.7407 in the test set.

4   Discussion

Our proposed framework was inspired by previous work of [3] and [20]. The main idea of this framework is to take a holistic perspective of intelligence predictions obtained from multiple ROIs. Initially, we attempted to feed the whole brain into the 3D ConvNet, and we attempted to train volumes of multiple ROIs while adding a concatenating layer on the top to merge the features from each ROI; these attempts, however, failed due to the memory constraints. In addition, we observed overfitting occurred very early, which was the reason including regularizer and the dropout layer within each repeated block unit, as well as training all the models for 3 epochs only.

Overall, the proposed framework showed that the median predicted scores from the left triangular part of inferior frontal gyrus, left caudate nucleus, and left thalamus contributed to a better prediction performance, compared to both the random guessing and to each region individually. Interestingly, the three ROIs all locate in the left hemisphere, and they have unique anatomical characteristics (see Fig. 2) and roles in cognitive functions. The triangular part of inferior frontal gyrus, also known as Brodmann area 45, is a cortical structure responsible for language processing [5]. Caudate nucleus is a subcortical structure and part of basal ganglia; it belongs to the corticostriatal circuitry and has many connections with frontal cortex and thalamus. This circuitry, particularly the caudate nucleus, contributes to goal-directed learning where the subject learns to recognize incentive perception to achieve a desirable outcome [7]. Thalamus is a subcortical structure located between the midbrain and cerebral cortex, with connections to many subcortical areas and the cerebral cortex. For a long time, thalamus was thought to be a hub that mainly relays information between different regions, but a recent study suggests that thalamus takes an active role in controlling functional cortical connectivity [17].

This research has several limitations. First, the SRI24 atlas [15] is derived from 24 participants spanning from late adolescence to late adulthood (age 19–84). Considering that the ABCD participants are 9–10 years old, the extent of age-related changes in cortical and subcortical structures can lead to biased (sometimes even misclassified) parcellations [4]. This is the primary concern that led us to refrain from interpreting the potentially biased anatomical changes, and we strongly suggest normalizing T1w data to an age-appropriate brain atlas to reduce biases and improve interpretability [4,6]. Second, we did not examine and exclude potential outliers in the datasets; utilizing the brain tissue segmentation volumetrics [13] provided by the organizer may help the process. Nevertheless, the strength of ConvNets is that the features learnt and extracted from images are much richer than the features (e.g., brain tissue segmentation volumetrics) derived from processing pipelines. Therefore, providing raw (or minimally processed) T1w images would be beneficial in extracting anatomical features for high-dimensional prediction tasks, as is previously shown [3]. Third, we trained the 3D ConvNets only for a few selected brain regions, and it is worthwhile to fully utilize the gray matter segmentations and explore whether other brain regions contribute to intelligence prediction.

In conclusion, our proposed framework suggested several cortical and subcortical brain regions that contribute to a better prediction of residualized fluid intelligence scores, compared to random guessing. Our framework can be validated and improved in the future, and it offers a new and unique perspective for predicting fluid intelligence based on brain morphometry.

Acknowledgement. We gratefully acknowledge Prof. Michael Zhu for facilitating centralized data storage, and Institute of Inflammation, Immunology and Infectious Disease for granting access to cluster computing resources provided by Information Technology at Purdue, West Lafayette, Indiana.

The data used in this report came from the ABCD Study Collection 3104 (https://nda.nih.gov/edit_collection.html?id=3104, accessed on or before March 24, 2019). Data access was in compliance with the NDA Data Use Certification and approved by the Institutional Review Board at Purdue University. The ABCD Study is supported by the National Institutes of Health and additional federal partners under award number U01DA041022, U01DA041028, U01DA041048, U01DA041089, U01DA041106, U01DA041117, U01DA041120, U01DA041134, U01DA041148, U01DA041156, U01DA041174, U24DA041123, U24DA041147, U01DA041093, and U01DA041025.

References

1. Akshoomoff, N., et al.: NIH toolbox cognition battery (CB): composite scores of crystallized, fluid, and overall cognition. Monographs of the Society for Research in Child Development (2013). https://doi.org/10.1111/mono.12038
2. Carroll, J.B.: Human Cognitive Abilities. Cambridge University Press, Cambridge (2009). https://doi.org/10.1017/cbo9780511571312
3. Cole, J.H., et al.: Predicting brain age with deep learning from raw imaging data results in a reliable and heritable biomarker. NeuroImage **163**, 115–124 (2017). https://doi.org/10.1016/j.neuroimage.2017.07.059
4. Dickie, D.A., et al.: Whole brain magnetic resonance image atlases: a systematic review of existing atlases and caveats for use in population imaging. Front. Neuroinformatics **11**, 1 (2017). https://doi.org/10.3389/fninf.2017.00001
5. Dronkers, N.F., Plaisant, O., Iba-Zizen, M.T., Cabanis, E.A.: Paul Broca's historic cases: high resolution MR imaging of the brains of Leborgne and Lelong. Brain **130**, 1432–1441 (2007). https://doi.org/10.1093/brain/awm042
6. Fonov, V., Evans, A.C., Botteron, K., Almli, C.R., McKinstry, R.C., Collins, D.L.: Unbiased average age-appropriate atlases for pediatric studies. NeuroImage **54**, 313–327 (2011). https://doi.org/10.1016/j.neuroimage.2010.07.033
7. Grahn, J.A., Parkinson, J.A., Owen, A.M.: The cognitive functions of the caudate nucleus. Prog. Neurobiol. **86**, 141–155 (2008). https://doi.org/10.1016/j.pneurobio.2008.09.004
8. Hagler, D.J., et al.: Image processing and analysis methods for the adolescent brain cognitive development study. bioRxiv p. 457739 (2018). https://doi.org/10.1101/457739
9. Ioffe, S., Szegedy, C.: Batch normalization: accelerating deep network training by reducing internal covariate shift. arXiv e-prints arXiv:1502.03167 (2015)
10. Jernigan, T.: Introduction. Developmental Cognitive Neuroscience (2018). https://doi.org/10.1016/j.dcn.2018.02.002
11. Kingma, D.P., Ba, J.L.: Adam: a method for stochastic gradient descent. In: ICLR: International Conference on Learning Representations (2015). https://arxiv.org/pdf/1412.6980.pdf
12. Li, S.C., Lindenberger, U., Hommel, B., Aschersleben, G., Prinz, W., Baltes, P.B.: Transformations in the couplings among intellectual abilities and constituent cognitive processes across the life span. Psychol. Sci. **15**, 155–163 (2004). https://doi.org/10.1111/j.0956-7976.2004.01503003.x

13. Pfefferbaum, A., et al.: Altered brain developmental trajectories in adolescents after initiating drinking. Am. J. Psychiatry **175**, 370–380 (2018). https://doi.org/10.1176/appi.ajp.2017.17040469

14. Plis, S.M., et al.: Deep learning for neuroimaging: a validation study. Front. Neurosci. **8**, 229 (2014). https://doi.org/10.3389/fnins.2014.00229

15. Rohlfing, T., Zahr, N.M., Sullivan, E.V., Pfefferbaum, A.: The SRI24 multichannel atlas of normal adult human brain structure. Hum. Brain Mapp. **31**, 798–819 (2010). https://doi.org/10.1002/hbm.20906

16. Rushton, J.P., Ankney, C.D.: Brain size and cognitive ability: correlations with age, sex, social class, and race. Psychon. Bull. Rev. **3**, 21–36 (1996). https://doi.org/10.3758/BF03210739

17. Schmitt, L.I., Wimmer, R.D., Nakajima, M., Happ, M., Mofakham, S., Halassa, M.M.: Thalamic amplification of cortical connectivity sustains attentional control. Nature **545**, 219 (2017). https://doi.org/10.1038/nature22073

18. Srivastava, N., Hinton, G., Krizhevsky, A., Sutskever, I., Salakhutdinov, R.: Dropout: a simple way to prevent neural networks from overfitting. J. Mach. Learn. Res. **15**, 1929–1958 (2014)

19. Tisdall, M.D., Hess, A.T., Reuter, M., Meintjes, E.M., Fischl, B., Van Der Kouwe, A.J.: Volumetric navigators for prospective motion correction and selective reacquisition in neuroanatomical MRI. Magn. Reson. Med. **68**, 389–399 (2012). https://doi.org/10.1002/mrm.23228

20. Wang, L., Wee, C.Y., Suk, H.I., Tang, X., Shen, D.: MRI-based intelligence quotient (IQ) estimation with sparse learning. PLoS One **10**, e0117295 (2015). https://doi.org/10.1371/journal.pone.0117295

21. White, N., et al.: PROMO: real-time prospective motion correction in MRI using image-based tracking. Magn. Reson. Med. **63**, 91–105 (2010). https://doi.org/10.1002/mrm.22176

Author Index

Printed in the United States
By Bookmasters